THE YALDON

PARTICLE
THEORY

The Explanation of Thermodynamics,
Propagation of Rays, Related Phenomena, and
the Model of the Atom

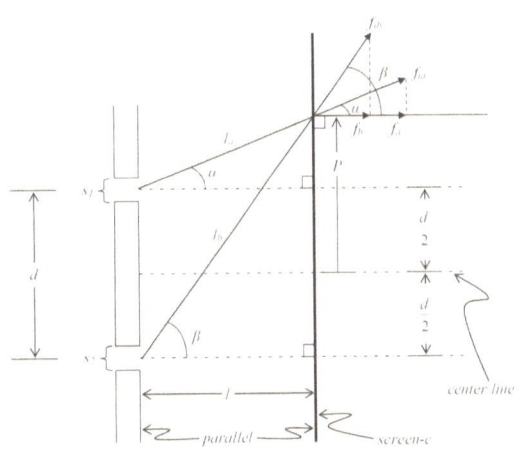

Menketh Yalda

Fulton Books, Inc.
Meadville, PA

Published by Fulton Books 2021

ISBN 978-1-63710-510-8 (paperback)
ISBN 978-1-63710-511-5 (digital)

Printed in the United States of America

I am thankful for the support of my wife, my children, and a special thanks to my youngest son for helping with some of the ideas of this theory.

CHAPTER 1

Introduction to Yaldon Particles and Thermodynamics

There are many unexplained phenomena and events that happen in our universe: black holes in the center of galaxies, giant pillars of elemental gasses in outer space, dark matter, the cause of gravitational force, earthquakes, volcanoes, the growth of stars in the universe, supernovas, and the redshift for the light that is received from far stars as well as the redshift for the light that leaves the earth. There are also other phenomena that one can experience: like heat, light, the electric and magnetic field, radio waves, microwaves, gamma rays, x-rays, and nuclear forces. Even though there are a lot of different forces and fields acting upon a person and their environment, all of these phenomena can be explained with the assumption of a byproduct of a constant influx of tiny and subatomic particles multi-directionally traveling throughout the universe. The subatomic particles can either travel individually or in groups through space. These subatomic particles will be referred to as *yaldons* for the purposes of further discussion, and to distinguish them from other tiny particle matter that have different characteristics from different theories.

A single yaldon particle has four main properties: average radius (r_y), average speed (s_y), average mass (m_y), and they are perfectly elastic (no loss in kinetic energy). With these four fundamental properties, yaldons govern the entire earthly and cosmic phenomena. A yaldon particle has no charge, spin, friction, internal structure, nor any other kinds of force fields surrounding it; these fields are the byproduct from the movement and arrangement of the yaldons. All the phenomena one experiences can be explained by using the four fundamental properties of yaldons. There will be no need to use models of ideas from science fiction: wormholes, time shrink, fabrics in space that stretch, or the big bang are a few examples of science fiction models that are used today.

Newton's second law of motion ($F\,t = m\,v$) will be the cornerstone for all calculations in this theory. According to Newton's second law, there cannot be momentum, force, energy, or time without a mass in relative motion to other points of references in a bounded system. The forces caused by fields (like the electric, magnetic, and gravitational field) are due to the change in momentum from the arrangement of yaldons, which constructs that particular field as it interacts with adjacent substances within a certain amount of time. All theories involving energy without the presence of a moving mass will not be included in the yaldon model since they go against Newton's second law of motion. It is assumed that yaldons are traveling linearly throughout the universe in all different directions, either traveling as a single particle or as a group of particles. Then the total net summation of their velocities in any region of space will not be zero but close to zero, due to the existence of objects (black holes, stars, planets, and elemental gases) in the universe. Each yaldon particle will also be assumed to have a spherical shape with an average radius of r_y, an average mass of m_y, an average speed of s_y, and they are perfectly elastic (loses no kinetic energy after each collision). These yaldons could have a different shape, but for purposes of calculations, each one will have an average spherical shape. The units of the metric system (*mks*) will be used for the purposes of calculations in the yaldon model.

The yaldons' density (ρ_y) is the number of traveling yaldons which move multi-directionally through an imaginary one square meter area plane (m^2) per one second of time (*sec*) from one side of that plane to the other side (fig. 1.1). Yaldons' density (ρ_y) is not the density of a single yaldon speck, but the total number of individual yaldons which pass through one side of a given square unit plane in one second of time. In other words, yaldons have an average speed (s_y) and travel in all different directions in space. Then the number count of yaldons that pass through an area plane of one square meter per one second in time from one side of that square meter area plane will be the yaldons' density (ρ_y) in that region of space (number of yaldons/m²sec) (see fig. 1.1).

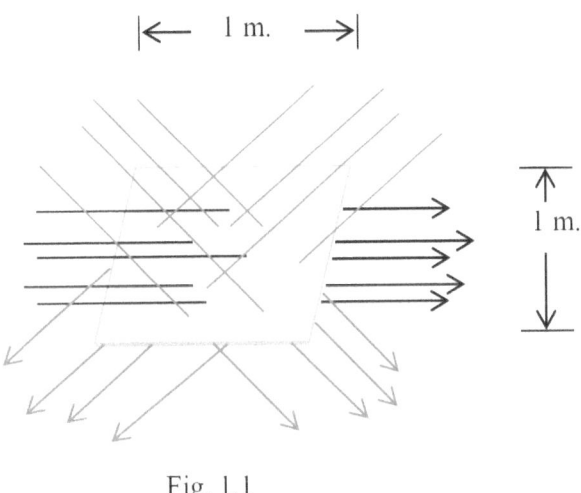

$$\text{|} \leftarrow \quad 1 \text{ m.} \quad \rightarrow \text{|}$$

Fig. 1.1

*A figure representing the yaldons' density (ρ_y) per one
second in time through one side of one square meter.

Whereas the density of a single yaldon speck (ρ_s) will be its mass (m_y) divided by the volume it occupies.
Then:

$$\rho_s = \frac{m_y}{\frac{4}{3}\pi \cdot r_y^3}$$

*Formula for the mass density of a single yaldon speck

7

As stated above, yaldons will either travel individually or in groups. The yaldons which travel in groups will have a density of ρ_g (*number of yaldon groups/m^2sec*). The average number of yaldons contained in one group of yaldon specks will be assumed to equal n_g. Then the total number of yaldons that pass through the one square meter area plane within one second of time by the combined density of ρ_y and ρ_g will equal ρ in empty space (far from any substance).

Then:

$$\rho = \rho_y + n_g \cdot \rho_g \ \ (N_y \, / \, (m^2 \cdot sec))\text{- - - - - } (1\text{-}1)$$

Yaldons travel linearly as a single particle or as a group of particles. The yaldons that travel individually through the universe in straight lines will be called *traveling yaldons*. They will be traveling as an individual yaldon speck, not as a group of yaldons. The yaldons that travel linearly in groups will be known as *traveling groups or traveling groups of yaldons*.

Calculating the Number of Yaldons in One Cubic Meter

As stated before, ρ is the number of yaldons (N_y) which pass through one side of a given plane of one square meter (*1 m^2*) within one second (*1 sec*) of time. As these yaldons pass through the given plane in one second, they will travel a certain distance from that plane which will be known as d (*distance in meters*). After they travel that distance (d) in one second, the number of yaldons that occupy the space from d to the given plane will be equal to ρ (the number of yaldons contained within an imaginary rectangular box [B] with a base of one square meter and a height of d, see following fig. 1.2 and 1.3).

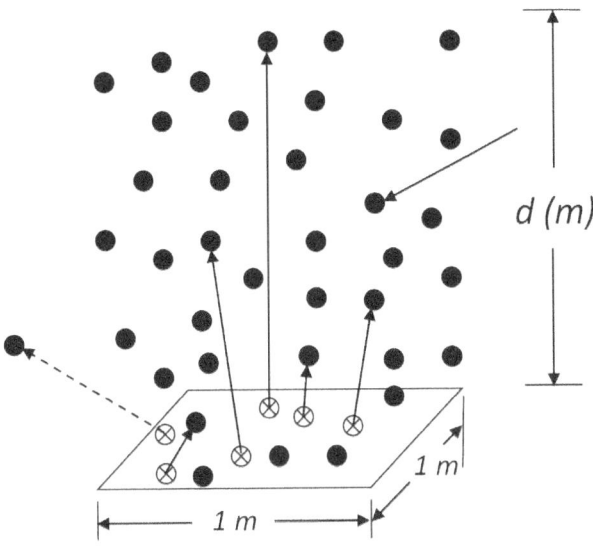

*Fig. 1.2 The number of yaldons which pass through a
one square meter plane within one second of time.

The distance (d) will be equal to the average speed (s_y) times one second:

d = (*average speed of yaldons*) × (*second*)

$$d = s_y \ \left(\frac{m}{sec} \right) \cdot 1 \left(sec \right)$$

Then:

$$d = s_y \left(m \right) \text{- - - - -} (1\text{-}2)$$

Let the volume of the rectangular box (B) be equal to V_B:

$$V_B = d \left(m \right) \times 1 \left(m \right) \times 1 \left(m \right)$$

After substituting the value of *d* from equation (1-2) above:

$$V_B = s_y (m) \times 1(m) \times 1(m)$$

Then:

$$V_B = s_y \left(m^3\right)$$

The number of yaldons (N_y) that entered the box (*Box B*) with a dimension of 1 (*m*) × 1 (*m*) × *d* (*m*) from the base (*Base [b]*) within one second of time will equal σ" (sigma double prime):

$$\sigma'' = \rho \times one\ square\ unit\ area \times one\ unit\ of\ time$$

$$\sigma'' = \rho \left[\frac{N_y}{(m^2 \cdot sec)} \right] \cdot 1(m^2) \cdot 1(sec)$$

Then:

$$\sigma'' = \rho \left(N_y\right)$$

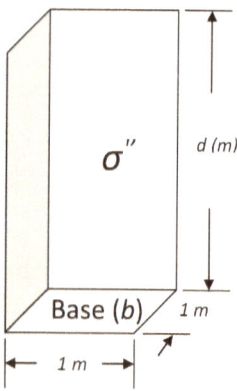

*Fig. 1.3 A rectangular box (B). This contains the number of yaldons that is equal to sigma double prime (σ") at any moment of time.

Let the number of yaldons that enter from one side of a one cubic meter ($1\ m^3$) box equal to σ' (sigma prime):

$$\sigma' = \frac{\sigma''}{V_B}$$

After substituting the values of σ'' and V_B:

$$\sigma' = \frac{\rho\left(N_y\right)}{S_y\left(m^3\right)}$$

Then:

$$\sigma' = \frac{\rho}{S_y}\ \left(\frac{N_y}{m^3}\right)$$

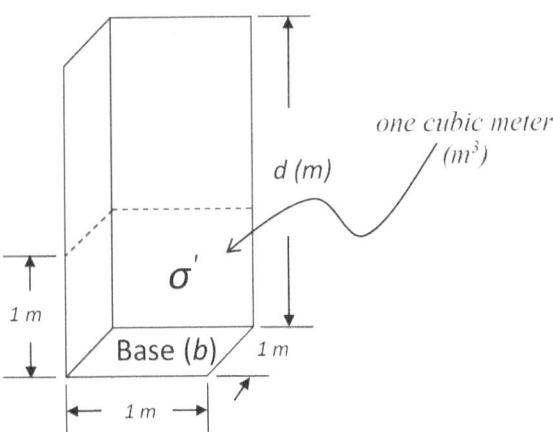

*Fig. 1.4 To calculate the number of yaldons contained within a one cubic meter box from one side of the cube, which is equal to sigma prime (σ').

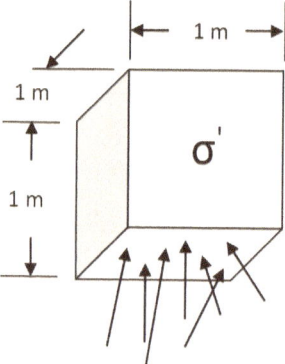

*Fig. 1.5 The number of yaldons that enter from one side of a cubic meter box is equal to sigma prime (σ') at any moment of time.

Sigma (σ) will equal the total number of yaldons that enter from all six sides of a one cubic meter box at any moment of time:

$$\sigma = 6 \cdot \sigma'$$

After substituting the value of σ':

$$\sigma = 6 \cdot \frac{\rho \left(\frac{N_y}{m^2 \cdot sec} \right)}{s_y \left(\frac{m}{sec} \right)}$$

Then:

$$\sigma = 6 \cdot \frac{\rho}{s_y} \left(\frac{N_y}{m^3} \right) \ \text{-----} \ (1\text{-}3)$$

During any instance in time, if a one cubic meter volume in outer space, far from any object, is taken and the momentum of each yaldon particle within that box are all added together (as a scalar value regardless of direction), then the total momentum of all the particles within that one cubic meter box will be M_y. Since the total number of yaldons that occupy a one cubic meter volume at any moment of time in outer space is σ, and the average speed of each yaldon is s_y, then:

$$M_y = \sigma \cdot m_y \cdot s_y \text{ - - - - - } (1\text{-}4)$$ Where m_y is the mass of a single yaldon particle.

Due to the perfect elasticity of a yaldon particle, the value of M_y will be conserved throughout the entire universe. In other words, if a one cubic meter volume is taken anywhere in the universe, even inside objects (including the atoms and molecules, except for black holes), then the total momentum for all the matter (yaldons, atoms, and molecules) within that cubic meter will still be equal to M_y. The average speed of yaldon particles inside objects will be less than s_y (s_y is the average speed of yaldons in empty space). Then the number of yaldons per cubic unit inside objects will be greater than the number of yaldons per cubic unit in empty space (σ). So the closer the atoms or molecules are, the greater the number of yaldons per cubic unit will become among the atoms or molecules of that substance. Overall, the substances that are more dense (atoms or molecules that are closer together with a slower rate of oscillation) will harbor a greater number of yaldons per cubic unit volume with a lower average speed in order to maintain the conservation of momentum (M_y) throughout the entire universe.

The conservation of momentum will force yaldon particles everywhere, in the space between the atoms and molecules of any object, in the universe. As previously mentioned, yaldons travel throughout the universe individually or in groups. The yaldons that travel in groups (traveling groups) will constantly strike the atoms and molecules of objects, causing them to vibrate back and forth around their rest points. This vibration will be governed by the law of simple harmonic motion. As atoms or molecules of an object are hit by traveling groups of yaldons, these atoms or molecules will vibrate along their rest points, ejecting groups of yaldons from the opposite side. As these groups of yaldons become ejected, there will be a vacant space that must be filled. The conservation of momentum will force individual yaldon particles to refill the space created from the ejected groups of yaldons. The refill of yaldon particles will usually provide a constant supply of yaldons for the atoms or molecules to constantly eject, as traveling groups continually strike the object. These ejected groups of yaldons will continue to travel inside the object and strike the next adjacent atom or molecule until all the atoms and molecules of that object will stay in constant vibration as the traveling groups of yaldons continually strike that object. The constant vibration, from all the atoms and molecules of an object, will cause the object to constantly release groups of yaldons. This constant release of propagated groups of yaldons from the object will be infrared and possibly other kinds of propagated rays.

All objects in the universe (not including black holes) will emit infrared at a steady state, far away from any source of extra external energy (for example, heat), due to the existence of traveling groups of yaldons, the conservation of momentum and simple harmonic motion. If an external source of heat is consistently applied to an object, the object's steady state will change. The external source of heat will increase the number of yaldon particles that reside between the atoms. This increase will result in a higher value for sigma (σ) in that object. A higher value for σ will cause the atoms of that object to propagate groups of yaldons with a higher momentum (increased temperature) than its steady state. These propagated groups with a higher momentum will cause the molecules of that object to become pushed farther away from each other than when in its steady state. Then the object will expand as the molecules become farther displaced, allowing a greater chance for the atoms and molecules in mixtures to react and form different compounds.

For example: as the molecules of a fuel (exposed to oxygen) become heated, they will react with the oxygen and produce new chemical compounds. As the fuel is heated, the number of yaldons around the fuel molecules (σ) will increase and force those molecules to become farther displaced as they oscillate. The larger displacement will cause the molecules in the fuel to vibrate with a greater distance (span) from rest point. This greater distance of oscillation will bring the fuel and oxygen molecules closer to one another, giving them a greater chance to react, as the traveling groups of yaldons bombard the fuel and oxygen molecules (see fig. 1.6). The fuel and oxygen, both, can harbor a certain amount of yaldon particles around their molecules. As the fuel and oxygen react, they will form new compounds. The new compounds formed from the chemical reaction will also harbor a certain amount of yaldon particles between their molecules. The certain amount of yaldons that the new compounds can harbor will be less than the amount of yaldons that the fuel and oxygen could harbor prior to the chemical reaction. Then there will be an extra amount of yaldons among the new compounds than the conservation of momentum will allow. This extra amount of yaldons will be propagated, as the traveling groups of yaldons bombard the molecules in the new compounds and released as extra heat. This release of extra heat will continue the chemical reaction between the fuel and oxygen until there is no more oxygen or fuel available. The continual release of heat from the new compounds formed is the flame of a fire.

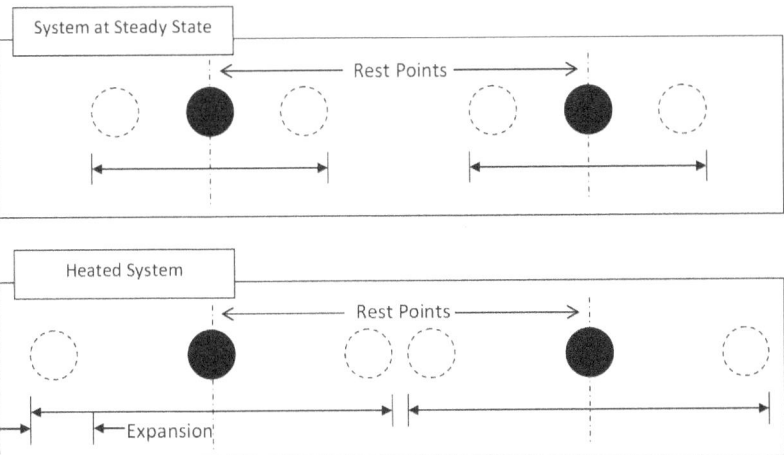

*Fig. 1.6 A rough diagram representing the expansion of an element after it is heated by an external source. The molecules of the heated system will come to a closer point to one another as they oscillate about their rest points.

In a similar way, sugar and oxygen molecules also harbor more yaldons than its new components (H_2O and CO_2). After being metabolized in a living body, the new components will have extra yaldons among them. The increase in the number of the yaldons around the new components (H_2O and CO_2) will raise the value of σ higher than the conservation of momentum will allow. As a result, the new components will radiate extra heat into the living body due to the bombardment from the traveling groups of yaldons.

Compressing gas molecules is an example to increase the value of σ more than the conservation of momentum will allow. As the molecules of the compressed gas are hit by the traveling groups of yaldons, they will propagate (radiate) the extra amount of yaldons in the form of propagated groups of infrared (heat) into the surrounding environment until the conservation of momentum (M_y) is maintained. After the conservation of momentum is maintained, the value of σ in the compressed gas will be greater when compared to the value of sigma in the non-compressed gas. Then there will be more yaldon particles per cubic unit among the molecules of the compressed gas, since the molecules in the compressed gas will be moving with a slower average speed (less momentum). In other words, the conservation of momentum will allow a greater number of yaldons to occupy the space among the molecules and atoms in the compressed gas. For example, four liters of gas at room temperature are compressed into one liter. After the conservation of momentum is maintained and the compressed gas has returned to room temperature, the value of the remaining σ in the one liter of compressed gas will be greater than the value of σ in one liter of non-compressed gas. Even though four liters of gas will have more yaldon particles among the molecules than the amount of yaldons in the same gas after compression. Friction is another example where the release of harbored yaldons among the molecules, as heat, is applied by removing the top layer of surface molecules in a substance.

CHAPTER 2

The Propagation of Rays

The vibration of atomic components, the orbital rotation of the atomic components, the vibration of molecules, and the vibration of the whole molecular structure each produce their own individual propagated ray. These propagated rays are visible light and other kinds of propagated rays (radio waves, microwaves, x-rays, gamma rays, etc.). The atoms and molecules propagate rays since they are surrounded by a swarming field of yaldons which move around the atom in all different directions (like bees buzzing around a hive). These yaldons are much smaller than any atomic component by far. The density (σ) of these swarming yaldons increases as it gets closer to the atomic components and molecules. As the atomic components and molecules vibrate back and forth through their rest points (each atomic component, molecule, and molecular structure will have its own rest point), they will eject a group of yaldons that are swarming around the atomic components and molecules. After each ejection from the oscillation of the atom, the swarming yaldons will fill in the space around the atom within a certain amount of time. The yaldons fill the space around the atom since they are under pressure from the conservation of momentum. The momentum of yaldons is conserved in space (M_y). The total momentum per cubic unit of yaldons, including the momentum of the atoms and/or molecules, will be the same anywhere in space. In this work, the term *molecular components* will be all the components in an atom, the atom as a whole, the molecules, and the molecular structure (in the case of solids). This oscillation from the molecular components causes the production of propagated rays. Following will be a rough diagram representing an atom and the yaldons that surround it, as well as the vibration of the atom and the ejection of propagated groups of yaldons.

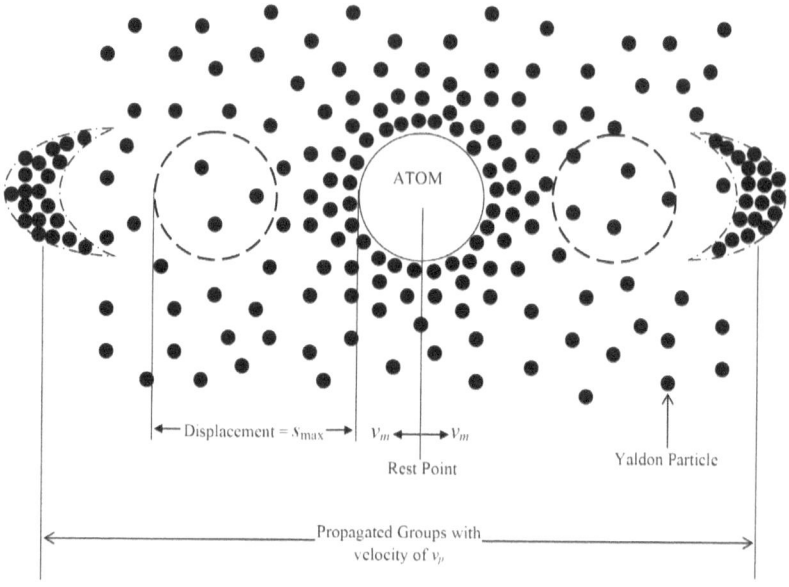

*Fig 2.1 Rough diagram representing the propagated group of yaldons emitted due to the vibration of atomic components in a substance. The vibration of molecules and molecular structures will be similar to this diagram.

All the atoms or molecules in substances are constantly vibrating and orbiting, releasing infrared, even though it may not have enough energy (momentum) to produce visible light. According to the figure above, as the atom in the center is displaced from its rest point, it will propagate a group of yaldons. The propagated ray is a constant series of groups of yaldons that have been launched by the oscillation of an atom. The yaldons in the propagated group will be consolidated into one cluster since all will have the same velocity (speed and direction). They will travel in a straight trajectory together as one whole group made of many yaldons. When the atom receives a form of energy from an external source, the amount of yaldons that swarm around the atom will also increase. As the atom oscillates, the number of yaldons contained in the propagated group will also increase since there is a larger amount of yaldons that swarm around the atom. The total mass of the propagated groups will be greater as more energy is received from the external source.

According to the formula of simple harmonic motion (SHM), there is a maximum displacement (s_{max}), a final (maximum) velocity from the oscillation of the atoms or molecules (v_m), and the mass of a molecular structure or its components (m) with a constant (k).

Simple harmonic motion formula:

$$v = \sqrt{\frac{k}{m}} \cdot s$$

$$v_m = \sqrt{\frac{k}{m}} \cdot s_{max} \text{-------- (2-1)}$$ *Where v_m is the final velocity of the oscillating molecule

As these atoms or molecules oscillate, they will launch propagated groups of yaldons with a final velocity of v_p. Since these propagated groups have a velocity and a mass, then they must also have momentum. The momentum of the propagated group depends on the final velocity (v_p) and the summation of all the yaldons' mass contained within that propagated group (Σm_y). The kind of ray emitted is determined by the momentum of several consecutive propagated groups that are produced within a certain time by the vibration of each molecular component contained within that substance. As a result, one substance will be able to produce more than one kind of propagated ray due to the variety of different molecular components within a substance (atomic components, the atom, the molecules, the molecular structure, the nuclei, etc.) as well as the concentration of the swarming yaldons around the atoms and molecules. Following are formulas and discussions based upon the propagated rays released from a substance. According to periodic simple harmonic motion, the frequency of the propagated groups from the oscillated atom or molecule is as follows:

$$f = \frac{1}{2\pi} \cdot \sqrt{\frac{k}{m}} \quad \text{-------- (2-2)}$$

The velocity of the propagated ray:

$$v_p = f \cdot \lambda \text{-------- (2-3)}$$

Then:

$$\lambda = \frac{v_p}{f} \quad \text{-------- (2-4)}$$

After substituting the value of $\sqrt{\dfrac{k}{m}}$ from equation (2-1) into equation (2-2):

$$f = \frac{1}{2\pi} \cdot \frac{v_m}{s_{max}}$$

Substituting the value of f from the equation above into equation (2-4):

$$\lambda = 2\pi \cdot s_{max} \cdot \frac{v_p}{v_m} \quad \text{-------- (2-5)}$$

Where λ is the distance between two consecutive propagated groups:

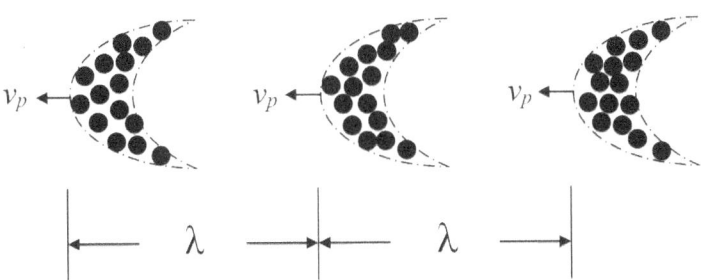

The propagated groups of yaldons, which are released from the vibration of the atomic component, will have a momentum with a frequency (wavelength) according to the energy received, the density of the swarming yaldons around the atomic component, and the atomic component's natural frequency. This model of propagated groups will satisfy and explain the phenomena of light's peculiar behavior as both a particle and a wave. The propagated groups behave as a wave since there is a periodic displacement between the consecutive groups, and they also have properties of a particle since the groups have a mass with a velocity (momentum). A conceptual example would be when a metal rod becomes heated by an external source of energy. The rod begins to emit the color red at first, and as the metal becomes hotter, it will begin to emit the colors that have a higher energy (higher momentum) than the red light. Energy and momentum are related since they both have a mass with a velocity (a moving mass or a mass in motion).

As the metal rod receives more heat (energy), its molecules will be surrounded by a greater number of yaldons that will swarm around the molecular components. The propagated groups will have a greater momentum with a frequency according to equation (2-2) and a wavelength according to equation (2-4) or (2-5). When the metal rod receives more heat energy from the external source, some of the molecular components on the surface will emit propagated groups with a greater number of yaldons, gradually increasing the mass of the ejected groups, and as a result, the momentum of the propagated groups of yaldons.

This growing momentum is seen when the metal rod first radiates red; as the momentum of the propagated groups grows the rod turns orange, then becomes yellow. This is caused by the molecular components being surrounded by a greater amount of swarming yaldons. As the metal rod continues to receive more energy, it releases the green, blue, and violet, along with the red, yellow, and orange, making the rod appear to emit white to the naked eye. This white-light effect is due to the entire visible spectrum of light being released from the surface molecules of the heated metal rod; according to the varying momentum of each propagated group from each molecular component, due to the different amount of yaldons contained within those propagated groups. The same phenomenon happens in the tungsten filament of a light bulb as it gradually becomes hotter, giving more momentum to the emitted groups of yaldons from the tungsten element until it begins to emit propagated groups of visible light. Following will be a diagram and formulas to find the relation between the energy received ($F \cdot s$) and the density of yaldons around the molecular components (σ) when at its steady state (the energy gained equals the energy released).

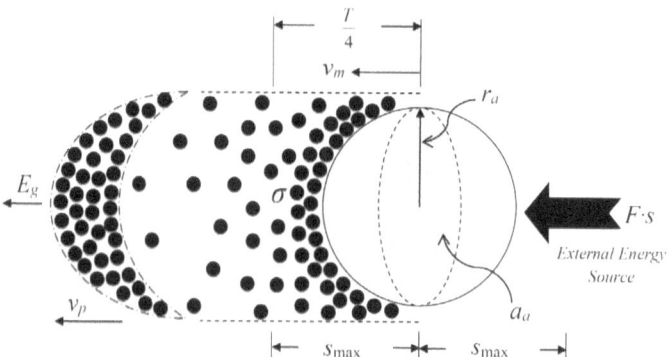

*Fig. 2.2 A rough diagram representing the relation between the energy of a propagated group of yaldons (Eg) and the density of yaldons around a molecular component (σ).

In the diagram above, σ (sigma) is the average number of yaldons per unit volume within a distance of $2s_{max}$. Following will be formulas to find the relation between the received energy from the external source and the energy of the propagated groups of yaldons (E_g).

According to Newton's second law:

$Ft = M_g$ (Where M_g is the momentum of one propagated group)

$$F = \frac{1}{t} \cdot M_g$$

After taking the derivative for the equation above:

$$dF = \frac{1}{t} \cdot dM_g$$

Let m_g be the total mass of the yaldons in one propagated group and v_p be the velocity of that propagated group. To find the total mass of the propagated group, the cross-sectional area of the molecular component (a_a) times the displacement of the molecule will give the volume of coverage. This volume times the average density of swarming yaldons around the atoms (σ) will give the number of yaldons that will be consolidated into a propagated group. The number of yaldons in the propagated group times the mass of a single yaldon (m_y) will give the mass of the propagated group (m_g).

Then:

$$m_g = a_a \cdot \sigma \cdot m_y \cdot s \quad \text{Where:} \quad a_a = \pi \cdot r_a^2$$

$$m_g = \pi \cdot r_a^2 \cdot \sigma \cdot m_y \cdot s$$

After taking the derivative:

$$dm_g = \pi \cdot r_a^2 \cdot \sigma \cdot m_y \cdot ds$$

And:

$$M_g = v_p \cdot m_g$$

Then:

$$dM_g = v_p \cdot dm_g$$

After substituting the value of dm_g into the previous equation:

$$dM_g = v_p \cdot \pi \cdot r_a^2 \cdot \sigma \cdot m_y \cdot ds$$

Substituting the value for dM_g into the equation for dF:

$$dF = \frac{1}{t} \cdot v_p \cdot \pi \cdot r_a^2 \cdot \sigma \cdot m_y \cdot ds$$

Where:

$t = \frac{T}{2}$ (Taking half of a period from the full cycle of oscillation $T=1/f$)

*A half of a period is taken since this will be considered as one full swing in one direction of the oscillating motion of the molecule.

From equation (2-2):

$$T = \frac{1}{f} = 2\pi \cdot \sqrt{\frac{m}{k}}$$

Then:

$$t = \pi \cdot \sqrt{\frac{m}{k}}$$

After substituting t into the equation for dF:

$$dF = \frac{1}{\pi} \cdot \sqrt{\frac{k}{m}} \cdot v_p \cdot \pi \cdot r_a^2 \cdot \sigma \cdot m_y \cdot ds$$

$$dF = \sqrt{\frac{k}{m}} \cdot v_p \cdot r_a^2 \cdot \sigma \cdot m_y \cdot ds$$

Performing integration:

$$F = \sqrt{\frac{k}{m}} \cdot v_p \cdot r_a^2 \cdot \sigma \cdot m_y \int_0^{2s_{max}} ds$$

$$F = 2\sqrt{\frac{k}{m}} \cdot v_p \cdot r_a^2 \cdot \sigma \cdot m_y \cdot s_{max}$$

Multiplying both sides of the above equation by $2s_{max}$:

$$2F \cdot s_{max} = 4\sqrt{\frac{k}{m}} \cdot v_p \cdot r_a^2 \cdot \sigma \cdot m_y \cdot s_{max}^2$$

After simplifying:

$$F \cdot s_{max} = 2\sqrt{\frac{k}{m}} \cdot v_p \cdot r_a^2 \cdot \sigma \cdot m_y \cdot s_{max}^2$$

After the excited system maintains equilibrium:

$$E_g = F \cdot s_{max}$$

Then:

$$E_g = 2\sqrt{\frac{k}{m}} \cdot v_p \cdot r_a^2 \cdot \sigma \cdot m_y \cdot s_{max}^2$$

Substituting the value of s_{max}^2 from equation (2-1) into the equation above:

$$E_g = 2\sqrt{\frac{m}{k}} \cdot m_y \cdot r_a^2 \cdot v_p \cdot v_m^2 \cdot \sigma \text{ -------- (2-6)}$$

And:

$$E_g = M_g \cdot v_p$$

After substituting E_g into equation (2-6):

$$M_g \cdot v_p = 2\sqrt{\frac{m}{k}} \cdot m_y \cdot r_a^2 \cdot v_p \cdot v_m^2 \cdot \sigma$$

Then:

$$M_g = 2\sqrt{\frac{m}{k}} \cdot m_y \cdot r_a^2 \cdot v_m^2 \cdot \sigma$$

According to equation (2-6), a higher density of swarming yaldons (σ) will result in a higher amount of energy that was received. Then highly dense substances, whose molecules have a larger radius (r_a), like the metal rod, will release propagated rays with a higher amount of energy. Whereas gasses under low pressure will have a lower value for σ and r_a, and will not be able to receive as much energy from an external source. Then gasses that are under a low pressure won't be able to emit a wide range of propagated rays. The low-pressure gases will be limited to propagate rays at a certain natural frequency, according to equation (2-2). If the same low-pressure gas becomes more condensed (a high-pressure gas), then the sigma will increase. This high-pressure gas will be able to receive a greater amount of energy from an external source and will then be able to propagate rays for a wide range.

These propagated groups of yaldons have two main characteristics. They are frequency and momentum within a certain amount of time. In the example regarding the visible light spectrum, which is detected by the human eye, the color for the visible light is due to the total momentum from the propagated groups of yaldons within a certain timeframe of t_p (see the following diagram). Propagated groups with a higher frequency (shorter λ) will have a greater number of groups within that certain timeframe. Neither the velocity, frequency, nor the total mass of a propagated group, but the summation of the total momentum of consecutive propagated groups within a certain timeframe (t_p) will designate color (as seen in equation '2-a' below). The brain will recognize a certain range of momentum from the propagated groups for each color as it strikes the eye nerves. This is the reason behind visible light "appearing" to have a defined set of colors. As seen in the previous example of the metal rod, visible light actually has a gradual blend of "colors" (momentum) from the propagated groups, but the brain partitions it into separate zones that are defined as red, orange, yellow, green, blue, indigo, and violet. This is similar to the way that the brain partitions musical notes (A, B, C, D, E, F, and G), even though the eardrum can vibrate with the frequencies in between these whole notes. The reason for visible light appearing to have a defined set of colors is due to the human brain, but in fact, the visible light and all other propagated rays are caused by the gradual variation (not step variation) of the momentum from the propagated groups, as seen in the example of the metal rod and tungsten filament.

The type of ray emitted depends on the total momentum from several propagated groups within a certain period of time. Following will be a diagram with formulas describing the energy from a series of propagated groups:

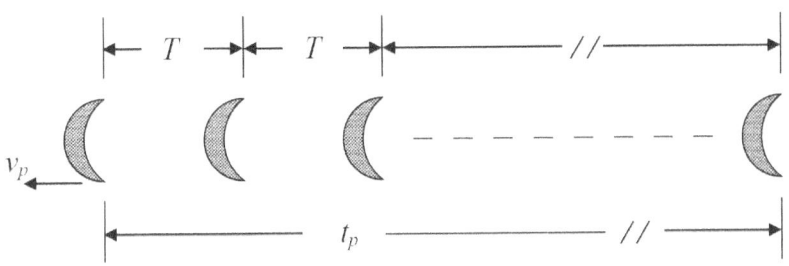

*A rough diagram representing several propagated groups
for a certain type of ray within a period of time (tp).

Let the total number of propagated groups within a certain timeframe (t_p) be N_g:

$$N_g = \frac{t_p}{T} \quad \text{Where: } T = \frac{1}{f}$$

Then:

$$N_g = f \cdot t_p$$

The momentum for one propagated group is M_g, and let the total momentum for all propagated groups within a certain timeframe (t_p) be M_p:

$$M_p = N_g \cdot M_g$$

After substituting the value for N_g into the previous equation:

$$M_p = M_g \cdot t_p \cdot f \text{-------- (2-a)}$$

Multiplying both sides of equation (2-a) by the velocity of propagated ray (v_p) to get a function for energy (E_p) within a timeframe of t_p:

$$\left(M_p = M_g \cdot t_p \cdot f \right) \times v_p$$

$$M_p \cdot v_p = \left(M_g \cdot v_p \cdot t_p \right) \cdot f \quad \text{Where: } M_p \cdot v_p = E_p$$

Then:

$$E_p = \left(M_g \cdot v_p \cdot t_p \right) \cdot f \text{-------- (2-b)}$$

Propagated Momentum-Shift Effect in Regard to Red Shift

The color of the visible light can have a *propagated momentum-shift effect* in empty space when the number of yaldons contained within the propagated groups change, but the velocity of the propagated groups will remain the same according to the law of inertia. For example, if the propagated groups of green light were to decrease in mass (less number of yaldons in the propagated group) while maintaining the same velocity, then it will become yellow light since less mass will mean less momentum (lower energy).

Empty space is filled with these yaldons, with less density (σ) in empty space than around matter. These yaldons which fill the empty space move around linearly in all directions with an average speed (s_y) that is faster than the velocity of the propagated visible light groups (the velocity of light). The yaldons that exist in empty space are the same yaldons that swarm around the atoms, and they are also the same yaldons that are contained in the ejected propagated groups.

As the propagated light groups of yaldons are released from the far stars in galaxies, they will come into contact with the yaldons that exist in space as they travel through. As these traveling propagated groups come into contact with the yaldons that exist in empty space, the propagated groups will lose some of their mass (yaldons) proportionally to the distance traveled while maintaining their same velocity. The yaldons that exist in empty space will knock off some of the yaldons in the propagated group, thus decreasing the propagated group's total mass while maintaining the same velocity. This decrease in momentum will cause a propagated momentum-shift effect. That is why the farther a galaxy is, the greater the shift in momentum. The redshift is caused by the existence of the yaldons in empty space, not by the universe expanding uniformly from any point of observation. The same effect happens to the propagated light groups which leave the Earth.

To find an equation that shows the ratio for the amount of momentum shift (R_{MS}) in outer space, the area of coverage from a propagated group (a_g) times the distance traveled will give a volume of space traversed by that propagated group. This volume multiplied by the density of the yaldons in outer space (σ) will give the total number of yaldons that come into contact with the propagated group. This total number times the cross-sectional area of one yaldon (πr_y^2), will give the total area of all the yaldons that come into contact with the yaldons in the propagated group that travels through space (a_s). Then the difference between the area of the propagated group (a_g) and the area of yaldons that come into contact with the propagated group (a_s), divided by the area of the propagated group (a_g) will give a ratio for the amount of the momentum-shift effect (redshift). Following will be a rough diagram with some formulas demonstrating this concept:

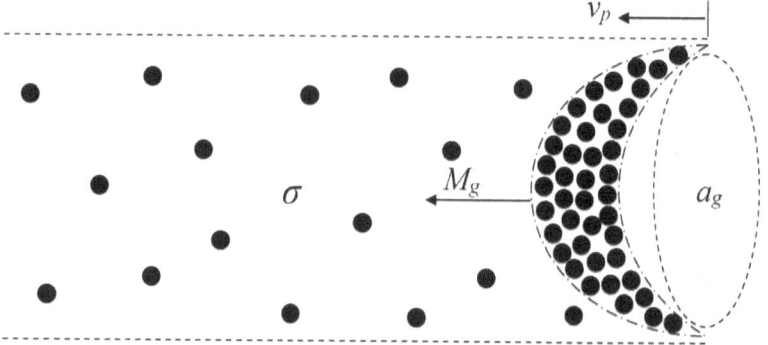

*Fig. 2.3 A rough diagram representing the volume covered by a propagated group as it travels through the density of sigma in space.

Finding the number of yaldons (N_{ly}) that will come into contact with the propagated group of yaldons for a distance of one light-year (O_{ly}):

$$N_{ly} = a_g \cdot O_{ly} \cdot \sigma$$

Let n_{ly} be the number of light-years that the propagated group travels, and a_s be the cross-sectional area for all the yaldons that come into contact with the propagated group during its travel through space:

$$a_s = \pi \cdot r_y^2 \cdot N_{ly} \cdot n_{ly}$$

The ratio for the momentum shift (R_{MS}) will be as follows:

$$R_{MS} = \frac{a_g - a_s}{a_g}$$

Then:

$$R_{MS} = 1 - \frac{a_s}{a_g}$$

Substituting the value for a_s in the equation above:

$$R_{MS} = 1 - \frac{\pi \cdot r_y^2 \cdot N_{ly} \cdot n_{ly}}{a_g}$$

Substituting the value for N_{ly} in the equation above:

$$R_{MS} = 1 - \frac{\pi \cdot r_y^2 \cdot a_g \cdot O_{ly} \cdot \sigma \cdot n_{ly}}{a_g}$$

Then:

$$R_{MS} = 1 - \pi \cdot r_y^2 \cdot O_{ly} \cdot n_{ly} \cdot \sigma \quad \text{-------- (2-7)}$$

When $R_{MS} \leq 0$, then it will not be possible to receive any kind of ray from that distance or beyond. According to the equation above, O_{ly} and r_y are fixed values. Then the amount of the momentum shift (redshift) for the propagated rays will depend on the number of light-years (n_{ly}) traveled. The more light-years traveled, then the greater the shift in momentum will be since they are directly proportional to each other. The sigma (σ) is also directly proportional to the ratio of the shift in momentum. Then the amount of the momentum shift will also increase based upon the density of the yaldons in that region. According to equation (2-7), the universe would appear to be spherical in shape with the observer at its center. The radius of this sphere would be from the observer to when the value of R_{MS} approaches zero.

Light's Velocity in Transparent Solid and Liquid Substances

When a propagated group of visible light travels through a transparent solid or liquid substance, it will strike the molecular structure, the molecules, the atoms, and the atomic components (molecular components) of that transparent substance. These solid and liquid transparent substances' molecular components are surrounded by a high density of swarming yaldons. Transparent gases under a low pressure will not have as high of a density of swarming yaldons as a transparent solid or liquid substance. This is due to gas molecules having a larger distance between one another, so their movement is not as restricted as a solid or liquid substance. In other words, gas molecules will have a greater velocity for v_m (as seen in fig. 2.4) than solid and liquid molecules. The way a propagated group of visible light travels through a transparent gas will be discussed later on in more detail since it will be slightly different than the way propagated light travels through a transparent solid or liquid.

As a propagated group of visible light strikes the surface of a transparent solid or liquid substance, the molecular components of that substance will shake a new propagated group from the opposite side of the initial strike. This new propagated group will have a greater mass than the propagated group from the initial strike and will then strike the next adjacent molecular structure or its components and so on, as it transfers through the entire transparent medium. These substances are transparent for visible light since the molecular components are able to transfer the same momentum that is delivered from the initial set of propagated strikes. The molecular structure or components, which cannot transfer the same momentum from the periodic strikes of propagated groups of visible light, will not be a transparent substance. Equation (2-10) will show whether or not a substance is transparent since it depends on the relation between the natural frequency of the molecules in a substance (f_m) and the frequency of the propagated ray (f_p) when $h = 1$.

According to this model of yaldon particles, the higher dense matter will harbor a higher density of the swarming yaldons that exist between the molecular components. The average number of the swarming yaldons that exist within the distance of $2s_{max}$ (as shown in fig. 2.4) per cubic meter will be known as sigma (σ). When a propagated group of visible light strikes the first molecular component in a transparent substance, the law of conservation of momentum will apply. As the momentum of the propagated group of visible light (M_g) strikes a molecular component in a transparent substance with a velocity of v_p, this will cause the molecular component to project a new propagated group of yaldons. This new propagated group will have a momentum (M_l) that is equal to M_g. The new propagated group will have a different velocity (v_l), depending on the mass of the molecular components in the transparent substance and sigma (number of yaldons that swarm around the molecules per cubic meter).

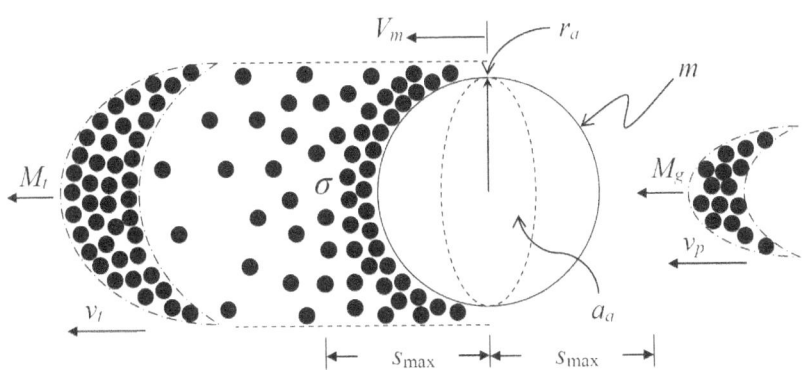

*Fig. 2.4 A rough diagram representing the incoming momentum (M¬g¬) striking an atomic component.

According to fig. 2.4, the molecular component will receive momentum from the incoming propagated group of yaldons (M_g) and will oscillate with a distance (span) that is equal to $2s_{max}$, with a maximum velocity equal to v_m. The molecular component will have a radius that is equal to r_a and a cross-sectional area that is equal to a_a. As the molecular component is displaced, it will propagate a new group of yaldons with a momentum that is equal to M_t. This new propagated group will have a mass (m_t) that is greater than the mass from the incoming propagated group, since the incoming group is traveling from a low-dense medium into a higher-dense medium. Then the velocity of this new propagated group (v_t) will be less than the incoming group since the molecular components of a transparent substance are able to retransmit the same momentum that is received from an incoming propagated group (M_g).

To find the mass of the new propagated group, the cross-sectional area of the molecular component times the displacement of the molecule ($2s_{max}$) will give the volume of coverage. This volume times the density (σ) will give the number of yaldons that will be consolidated into the new propagated group. The number of yaldons in the new propagated group times the mass of a single yaldon (m_y) will give the mass of the new propagated group (m_t).

Then:

$$m_t = a_a \cdot \sigma \cdot m_y \cdot s \quad \text{Where: } a_a = \pi \cdot r_a^2$$

$$m_t = \pi \cdot r_a^2 \cdot \sigma \cdot m_y \cdot s$$

After taking the derivative of the equation above:

$$dm_t = \pi \cdot r_a^2 \cdot \sigma \cdot m_y \cdot ds$$

The momentum of the new propagated ray (M_t) will be the mass (m_t) times the velocity (v_t):

$$M_t = v_t \cdot m_t$$

After taking the derivative for the equation above:

$$dM_t = v_t \cdot dm_t$$

After substituting the value of dm_t into the equation above:

$$dM_t = \pi \cdot v_t \cdot r_a^2 \cdot \sigma \cdot m_y \cdot ds$$

Performing integration:

$$\int dM_t = \pi \cdot v_t \cdot r_a^2 \cdot \sigma \cdot m_y \int_0^{2s_{max}} ds$$

$$M_t = 2\pi \cdot v_t \cdot r_a^2 \cdot \sigma \cdot m_y \cdot s_{max}$$

Since momentum is conserved:

$$M_g = M_t$$

Then:

$$M_g = 2\pi \cdot v_t \cdot r_a^2 \cdot \sigma \cdot m_y \cdot s_{max}$$

After solving for v_t:

$$v_t = \frac{M_g}{2\pi \cdot r_a^2 \cdot \sigma \cdot m_y \cdot s_{max}} \quad - - - - - - - - (2\text{-}8)$$

Where: $M_g = m_g v_p$

After substituting the above equation into equation (2-8):

$$v_t = v_p \cdot \frac{m_g}{2\pi \cdot r_a^2 \cdot \sigma \cdot m_y \cdot s_{max}}$$

Substituting the value for s_{max} from equation (2-1) into the equation above:

$$v_t = \frac{v_p}{v_m} \cdot \frac{m_g}{2\pi \cdot r_a^2 \cdot \sigma \cdot m_y} \cdot \sqrt{\frac{k}{m}} \quad\text{------- (2-9)}$$

As one can see from equation (2-9), the velocity of light in a transparent substance (v_t) will be inversely proportional to the mass (m) and the radius (r_a) of the molecular component in the transparent medium. As a result, the larger a molecular structure is, the slower the light will be as it travels through that transparent medium. Also, the density (σ) is inversely proportional to the velocity. Then the larger groups of yaldons that are propagated from dense transparent materials will travel slower, but with more mass. This effect is seen when light travels through a diamond, compared to when light travels through glass. The diamond is a more dense substance than glass. Then the propagated groups of light that travel through a diamond will move at a slower velocity when compared to the propagated groups of light that move through the glass.

The same effect applies when light leaves the highly dense transparent substance and returns to the less dense transparent medium. The molecules on the surface of the less dense transparent medium will have a less density of swarming yaldons that surround it. Then the less dense transparent medium will transfer the propagated visible light group of yaldons with a higher velocity, but with less mass (refer to equation 2-9). This is the reason behind light regaining its original velocity as it leaves the glass and goes back into the air.

Emission and Absorption Lines in Low Pressure Gasses

The molecules and atoms of gasses under a low pressure are surrounded by a low density of swarming yaldons (σ). When the molecules, atoms, and atomic components in a low-pressure gas receive an external form of energy; each of those molecular components will increase the span of oscillation while maintaining its own natural frequency. The molecules in a low-pressure gas are able to freely oscillate about their rest points with a greater span than a high-pressure gas would; since high-pressure gas molecules are more constricted by the limited space between the molecules, as well as the high density of swarming yaldons between the high-pressure gas molecules. The density of yaldons that swarm around the atoms and molecules will be greater as the atoms and molecules become closer together. When the density of yaldons that swarm around the molecules becomes greater, in the case of high-pressure gasses, then the increase in the density of yaldons around the molecules (σ) will cause a high gain in the momentum for the propagated groups. The opposite is also true, the density of swarming yaldons will be lower when the atoms and molecules are farther apart from each other. Then the large displacement of molecular components in a low-pressure gas, combined with the low density of swarming yaldons, will result in a periodic ejection of propagated groups of yaldons with a consistent momentum. This consistent momentum will produce an emission line of a single color for each molecular component in the low-pressure gas.

High-pressure gases, which have a higher amount of swarming yaldons between the molecules, will release more energy. The propagated groups of yaldons produced from a high-pressure gas will be more massive than a low-pressure gas. If the propagated groups from a high-pressure gas were to receive more external energy, then the change in momentum would be greater than if a propagated group from a low-pressure gas were to receive more external energy. As a result, the high-pressure gas will be able to release propagated rays with a broader range of momentum when compared to a low-pressure gas. This broader range of momentum will allow the high-pressure gas to produce a fuller spectrum of color when the molecules are excited by an external source of energy.

As mentioned before, all the atomic components and molecules in a low-pressure gas will release a group of yaldons with a consistent momentum, and each molecular component will emit a certain color when it receives energy from an external source. Then the colors of the emission lines will depend on the momentum produced from the natural oscillation of each atomic component and molecules in the low-pressure gas, due to the fixed period and span of oscillation from the atoms and molecules. Larger elements, which have more components in their atoms, will have more emission lines produced. The emission lines are caused by every atomic component and molecule in the low-pressure gas having its own natural frequency and periodically propagating rays at a consistent momentum. This is shown in equation (2-6), when σ has a low value then the substance cannot produce a wide range of propagated rays.

When the emission lines produced by the excited state of the low-pressure gas is passed through the same kind of low-pressure gas but at a lower temperature (colder), the emission lines that are produced by the natural oscillation of the molecules and atoms in the excited low-pressure gas will not be shown as it passes through the cold low-pressure gas. The molecules and atoms in the cold low-pressure gas will resonate with the emission lines by having a greater displacement from their rest points. This resonance will cause the span of oscillation of the molecules and atoms in the cold low-pressure gas to become greater with every consecutive strike from the propagated groups of yaldons in the emission lines produced by the same gas at an excited state. As the span of oscillation of the molecules and atoms in the cold low-pressure gas becomes greater, the velocity of the molecules and atoms will also increase, to a point where the molecules and atoms will be forced to leave their rest points. Then the molecules of the cold low-pressure gas will not be able to propagate groups of yaldons at the specific colors which resonate with the natural oscillation of the molecules and atoms. The other propagated groups of light that do not match the natural frequency of the molecules and atoms in the cold low-pressure gas will be able to transfer through since they do not resonate with the natural frequency of those molecules (see equation 2-10). A radiometer is another example of this effect since the fans in the radiometer will spin due to the increased range of motion from the atoms and molecules in the low-pressure gas as they oscillate and leave their rest points.

This effect of resonance between the external light source, and the oscillation of the atoms and molecules can be compared to the resonance between the molecules in a glass cup and a certain frequency of sound. The molecules in the glass cup will have a certain natural frequency based upon its molecular structure and shape of the cup. When an external sound wave that matches the natural frequency of the cup is applied, the resonant frequency from the sound wave causes the molecules in the glass to become displaced to a point where they can no longer maintain their position in the structure of the glass cup. As a result, the glass cup shatters. The same effect happens to the oscillation of the molecules and atoms in a cold low-pressure gas when given an external form of light that matches the natural frequency from the oscillation of the molecules and atoms. The molecules and atoms become so displaced from their rest points, they will not be able to produce a continuous stream of propagated groups of yaldons for that specific frequency of light. As a result, the oscillation of the molecules and atoms which resonate with that specific color becomes "shattered." Following will be a diagram and formulas to show the relation between the natural frequency of a substance (f_m) and the frequency of the incoming propagated groups (f_p), the density of the swarming yaldons between the molecules and atoms (σ), and the velocity of the molecules and atoms (v_{mh}):

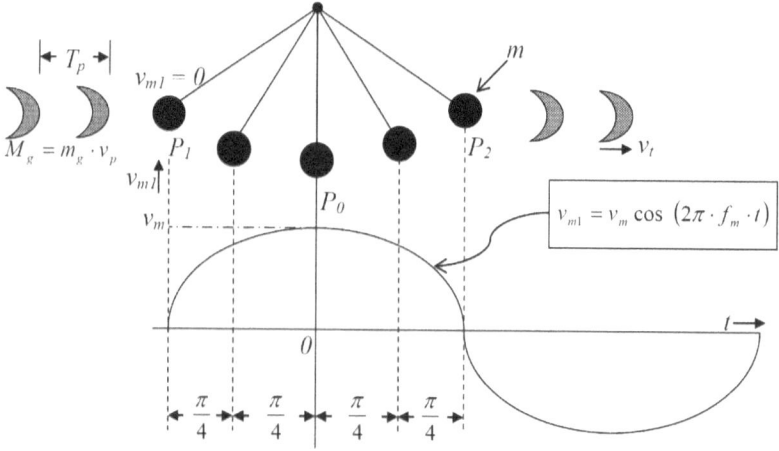

*Fig. 2.5 A rough diagram representing the oscillation of an atom with periodic strikes from propagated groups.

In the figure above, the oscillation of the atom is represented by the motion of a pendulum to make the concept of oscillation easier to understand. In order for resonance to happen, each of the propagated groups that travel with a period of T_p should strike the mass, m, at the position of P_1 when $v_{m1} = 0$ (for max. resonance). The equation for v_{m1}, in the diagram above is a sinusoidal representation of the velocity for the mass of the atom, m, with its natural frequency of f_m, and v_m will be the maximum velocity of that atom after the first strike from a propagated ray. The value for v_m will come from equation (2-9).

According to equation (2-9):

$$v_m = \frac{v_p}{v_t} \cdot \frac{m_g}{2\pi \cdot r_a^2 \cdot \sigma \cdot m_y} \cdot \sqrt{\frac{k}{m}}$$

v_p is the velocity of light in a vacuum, v_t will be the velocity of light in a transparent substance, and v_m will be the maximum velocity of the oscillating molecule. The value of v_m will be the amplitude for the equation $v_{m1} = v_m \cos(2\pi \cdot f_m \cdot t)$, after the first strike from the incoming propagated groups. The equation below will be for the second strike from the incoming propagated groups, which will have a frequency of f_p:

$$v_{m2} = v_{m1} + v_m \cos\left(2\pi \cdot f_{m1} \cdot \left(t + \frac{1}{f_p}\right)\right) \quad \text{Where: } T_p = \frac{1}{f_p}$$

For n number of hits from the incoming propagated groups within a certain period of time, the equation above can be rewritten as follows:

$$v_{mh}(t) = \sum_{n=0}^{h} v_{mn}$$

Then:

$$v_{mh}(t) = \sum_{n=0}^{h} v_m \cos\left(2\pi \cdot f_m \cdot \left(t + \frac{n}{f_p}\right)\right)$$

After substituting the value for v_m in the equation above:

$$v_{mh}(t) = \sum_{n=0}^{h} \frac{v_p}{v_t} \cdot \frac{m_g}{2\pi \cdot r_a^2 \cdot \sigma \cdot m_y} \cdot \sqrt{\frac{k}{m}} \cdot \cos\left(2\pi \cdot f_m \cdot \left(t + \frac{n}{f_p}\right)\right)$$

After rewriting the previous equation:

$$v_{mh}(t) = \frac{v_p}{v_t} \cdot \frac{m_g}{2\pi \cdot r_a^2 \cdot \sigma \cdot m_y} \cdot \sqrt{\frac{k}{m}} \left[\sum_{n=0}^{h} \cos\left(2\pi \cdot f_m \cdot \left(t + \frac{n}{f_p}\right)\right)\right] \quad \text{-----} (2\text{-}10)$$

According to equation (2-10), the velocity from the oscillation of the atoms and molecules (v_m) will sharply increase as the incoming propagated strikes match the frequency of resonance ($f_p=f_m$). As the value of h (number of consecutive hits) increases, the maximum velocity from the oscillation of the atom and molecules will increase linearly. Within a fraction of a second, there will be numerous amounts of consecutive hits from the propagated groups. This will escalate the velocity from the oscillation of the atoms and molecules (v_{mh}) to a value where the atoms and molecules can no longer keep their original rest points (when the glass cup shatters by a resonant frequency). This is why cold low-pressure gases cannot continuously propagate rays at those specific frequencies of resonance, since the small value for σ will cause the gas to radiate less energy than what was received from the oncoming emission lines (see equation 2-6). As a result, the frequencies from the propagated rays which match the natural frequencies of the cold low-pressure gas will be absorbed by that gas.

The Constant Velocity of Propagated Rays

All atoms and molecules are the basic building blocks for anybody in the universe (except for black holes which are a special case) and they are surrounded by swarming yaldons. If a body in the universe moves with a velocity (v_B), then the summation of the average relative net velocity of the swarming yaldons around the atoms and molecules of that body will be a negative value (opposite direction) for v_B. When a body in the universe is not in motion, then the summation of the average relative net velocity of the swarming yaldons around the atoms or molecules of that body will be close to zero. Following will be a diagram to illustrate this concept.

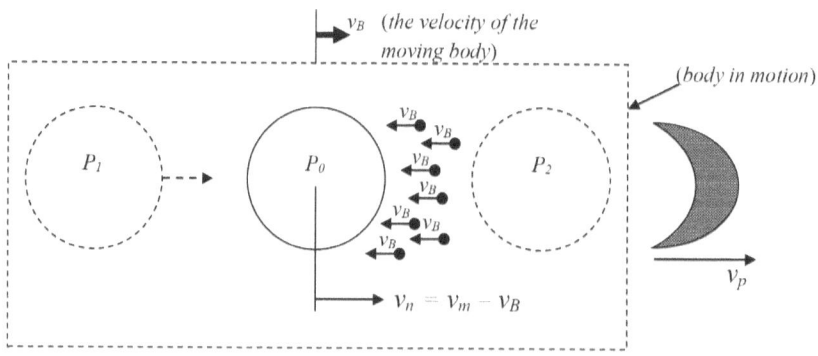

*Fig. 2.6 A rough diagram representing the propagated ray group maintaining its velocity (v_p). The net velocity is v_n for the atom or molecule while it is oscillating and in motion.

The velocity of the moving body is v_B. When the atom moves from P_1 to the position of P_0, the velocity of the atom will be v_m. Since the body is in motion with a velocity of v_B, the net velocity (v_n) for the atom at the position of P_0 will be $v_n = v_m - v_B$. The final velocity of the atom will be v_f, since the whole atom will move with a velocity of v_B (the velocity of the moving body) plus the velocity of v_n.

Then:

$$v_f = v_B + v_n$$

Where: $v_n = v_m - v_B$

$$v_f = v_B + \left(v_m - v_B \right)$$

$$v_f = v_m$$

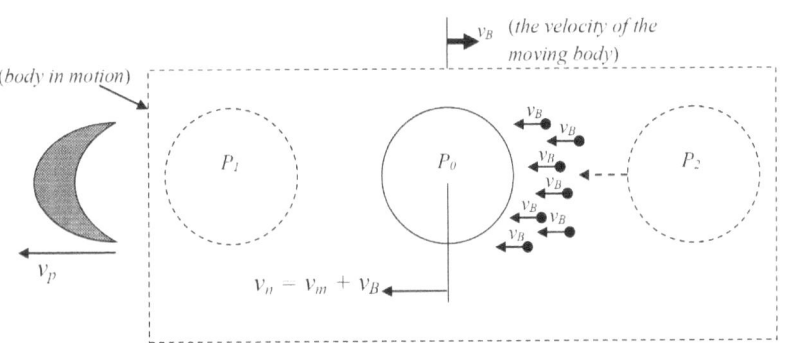

*Fig. 2.7 A rough diagram representing the propagated ray group from the opposite side also maintaining its velocity (v_p).

The atom that is illustrated in fig. 2.7 is contained in a body that is moving with the same velocity (speed and direction) of v_B from fig. 2.6. The difference will be the direction of oscillation from P_2 toward P_0. The net velocity (v_n) at position P_0 will be

$v_m + v_B$ since the resultant net velocity from the swarming yal-dons around the atom will be v_B. The resultant net velocity of the swarming yaldons will be in the same direction of the atom's movement at position P_0, but the whole body (which contains the atom) is moving in the opposite direction of the net velocity. Then the final velocity (v_f) of the atom at the position of P_0 will be:

$$v_f = v_n - v_B$$

The net velocity will be:

$$v_n = v_m + v_B$$

Then:

$$v_f = (v_m + v_B) - v_B$$

$$v_f = v_m$$

According to the diagrams and formulas above, the veloc-ity of the propagated rays (velocity of the light) will be constant, regardless of the velocity of the body that emits the propagated ray (source of the light). The same principle applies to a stationary light source, but the observer is in motion. According to equation (2-9) rewritten for v_p:

$$v_p = v_m \cdot \frac{v_t}{m_g} \cdot 2\pi \cdot r_a^2 \cdot \sigma \cdot m_y \cdot \sqrt{\frac{m}{k}}$$

According to the equation above, as long as the maximum velocity of the atom (v_m) remains the same, then the velocity of the propagated groups will also remain constant. These are not the only phenomena that can be explained by using yaldon propaga-tion. For example, the refraction, diffraction, interference, disper-sion, and polarization of light, as well as many other phenomena, can be sufficiently understood using this model.

CHAPTER 3

Explanation of Phenomena Related to the Propagation of Rays

The Force Caused by the Traveling Groups of Yaldons

Normally, the forces are balanced around all the molecules or atoms in a substance with a rectangular shape (excluding the corners). This is due to the incoming bombardment from the traveling groups of yaldons onto all sides of that substance. A location where the resultant forces are different in a substance is on the surface molecules of a liquid or solid substance. For example: if a solid substance is placed in an environment with air (gas), then the molecules on the surface of the solid substance will oscillate with a greater displacement than the molecules below the surface. Following will be fig. 3.1 representing the forces between the molecules with springs. In fact, there are no springs between the molecules of any substance, but the nature of the forces between the molecules and/or atoms behaves as a spring.

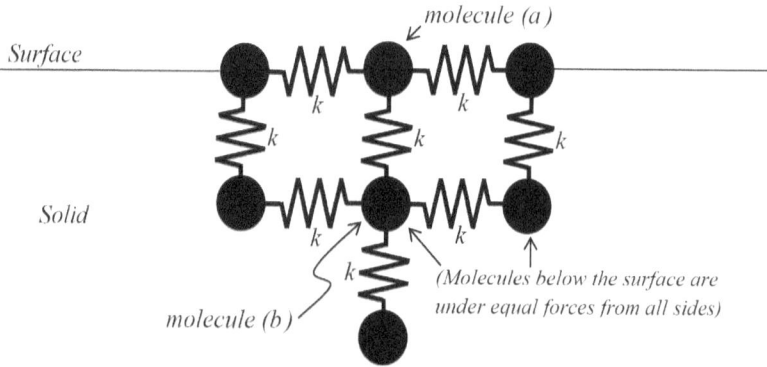

*Fig. 3.1 A rough diagram representing the forces between the molecules. The molecules below the surface of the rectangular solid will have balanced forces acting upon them. The forces between them are represented by springs with the constant k.

When any substance is placed in a less dense medium, as in fig. 3.2, the traveling groups will strike the surface molecules of the more dense substance from all directions with varying angles. The area (a_s) will represent the cross-sectional area of one molecule on the surface of the more dense substance. Then that cross-sectional area (a_s) will be under a constant bombardment from the traveling groups with different angles of α (in respect to the y-axis), and each hit from the traveling groups will apply a force that can be analyzed into two components $(f_{pxz}$ and $f_{py})$. The value of the two components will depend upon the angle of incidence (α) and the momentum of the incoming group. Following will be fig 3.2, which will illustrate the resultant force (f_{py}) applied perpendicularly onto a surface molecule (*molecule (a)*) from the continuous bombardment of traveling groups:

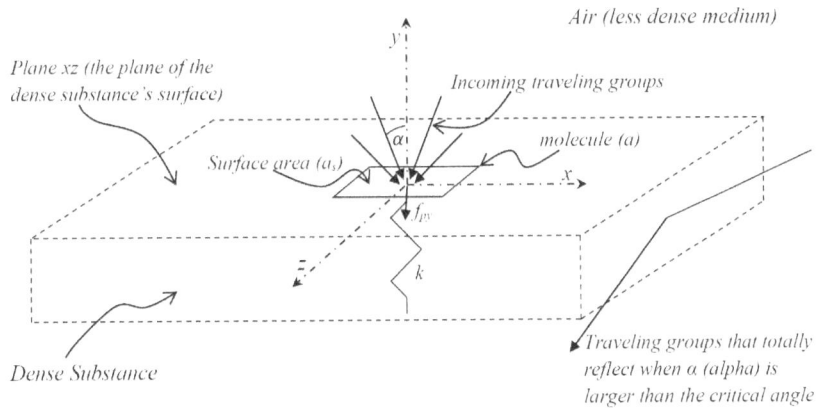

*Fig. 3.2 A rough diagram representing the resultant force (f_{py}) on the surface molecules from the incoming traveling groups.

The component f_{pxz} will cancel each other out, since the traveling groups will hit the cross-sectional area (a_s) from different directions and varying angles in respect to the y-axis, as seen in fig. 3.2. As a result, the only component that will act on the surface molecules is the force f_{py}, and this force causes the surface molecules to vibrate vertically along the y-axis. An example to relate this concept to is a trampoline. The components f_{pxz} can be compared to the springs which hold the trampoline taut in the horizontal direction. As a result, the surface of the trampoline will only be able to move in a vertical direction. As well as oscillating vertically in the direction of the y-axis, surface molecules will also have a larger span of oscillation than the rest of the molecules that are internally within the dense substance. Following will be fig. 3.3, which will represent the cancellation of the forces, f_{pxz}, on *molecule (a)* as the traveling groups bombard it from different directions and angles.

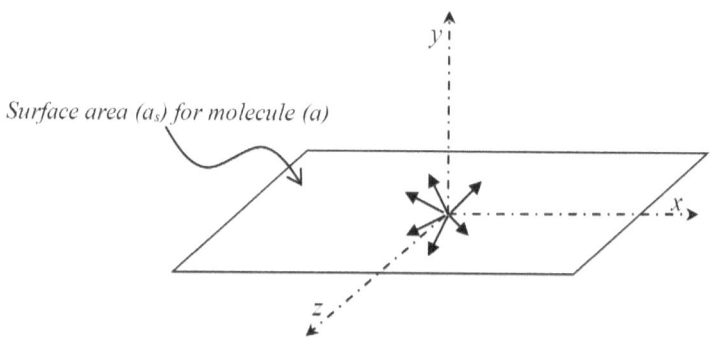

*Fig. 3.3 A rough diagram showing the components of f_{pxz}, applied to *molecule (a)* on the *Plane xz*, due to the strikes from several traveling groups onto that surface molecule.

The traveling groups also strike *molecule (a)*, seen in fig. 3.2, from the opposite side (the side of *Plane xz* that is toward the internal structure of the dense substance), but the rate of the traveling groups coming from that side will be less than the rate of traveling groups which come from the less dense medium. This is due to the more dense substance totally reflecting the traveling groups that strike the surface at an incident angle (α) which is more than the critical angle (see fig. 3.2). Then the force, f_{py}, is a resultant force from the bombardment of traveling groups onto both sides of the surface *molecule (a)*. The critical angle of reflection does not only apply to visible light but also to traveling groups of yaldons. The differences are that traveling groups of yaldons do not have a periodic displacement between them, and each group will travel in a non-uniformed (random) direction. Thus, the traveling groups of yaldons will not be considered as a propagated ray.

Larger molecular structures will be under a higher rate of strikes from the traveling groups, since they will have a larger cross-sectional area. As a result, there will be a stronger resultant force of f_{py} $\left(f_{py} \propto r_a^2 \right)$. Then the surface molecules of that substance will have a greater span of oscillation when placed in a less dense medium. For example, the molecular structure of a diamond is larger than glass. Then the surface molecules of a diamond will travel at a greater distance from the surface as it oscillates when compared to the distance that the surface molecules of a glass travel. This greater displacement of the surface molecules in a diamond will give it the characteristic of being able to scratch the surfaces of other solid substances. Also, this vertical oscillation of surface molecules caused by the resultant force of f_{py} from the bombardment of traveling groups will not allow a broken substance to be placed back together. Tearing a sheet of paper in half is an example of this effect, since molecules on the new surfaces (where the tear happened) will swing farther than they had prior to being torn apart. The span of oscillation on the new surfaces will be too great when compared to the small distance required to maintain the forces between its molecular structures. As a result, the surface molecules of the sheet of paper will no longer be able to be placed back together again after being torn apart. Breaking a piece of glass into two parts is another example of the surface molecules oscillating at a larger distance (greater span). The two pieces of glass will no longer be able to be placed back together due to the effect from the larger span of oscillation produced by the molecules on the new surfaces.

Total Reflection

The forces between the atoms or molecules of any substance can be represented by springs with different values for the spring constant k. The value of k will depend on type of substance or element; different substances will have different values for the spring constant k (see fig. 3.1). When one propagated group of light strikes the *surface area* (a_y), as in the following fig. 3.4, with a force of f_i; f_i can be analyzed into two components (f_{ix} and f_{iy}).

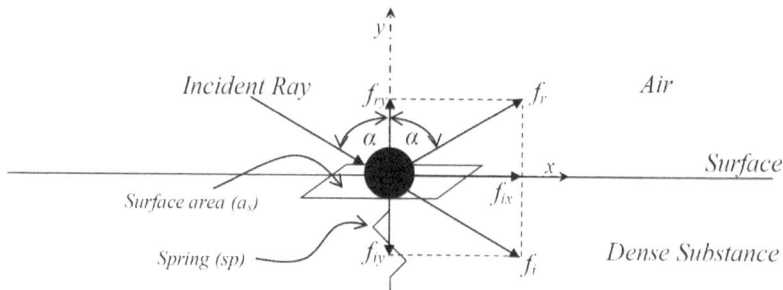

*Fig. 3.4 A rough diagram representing the components of forces acting upon the surface molecule as a propagated group of light strikes it.

The component force, f_{iy}, will compress *spring (sp)*, as in the figure above. There will be a distance of compression (s) and a required rate of time for the *spring (sp)* to store the kinetic energy from the impulse of the incident ray group as potential energy. When the *spring (sp)* expands, it will apply a force equal to f_{iy}, but in the opposite direction (at an angle of 180°), as seen in the figure above with the force, f_{ry}. This is due to the surface molecules oscillating in the vertical direction, in regard to that surface. Then the incident ray will reflect in the direction of the force, f_r, at an angle equal to α in respect to the *y-axis*, as seen in the figure above. This force (f_r) is a resultant force from the components f_{ry} and f_{ix}. This will result in a total reflection for the propagated ray (incident ray). The kinetic energy (E_{ki}) for one propagated group from the incident ray is:

$$E_{ki} = \frac{1}{2} m_g v_p^2$$

Where m_g is the total mass of yaldons contained in one propagated group. In order to have total reflection on the surface of any substance, the value of the constant k must be large enough to store the kinetic energy from one propagated group of the incident ray. Following will be a diagram and formulas modeling this concept.

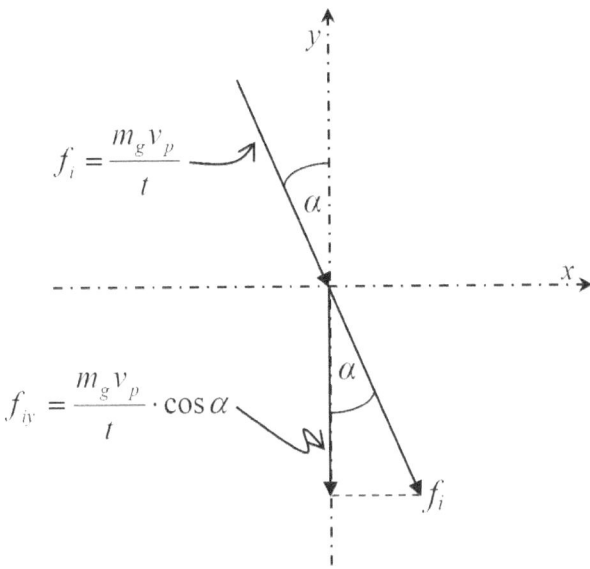

*Fig. 3.5 A rough diagram showing the relation between the force and momentum of one propagated group from the incident ray.

$$f_{iy} = \frac{m_g v_p}{t} \cos \alpha$$

$$f_{iy} = f_i \cos \alpha$$

$$f_{iy} = ks \qquad \text{Hooke's law}$$

$$f_{iy} \cdot ds = ks \cdot ds \qquad \text{Multiply both sides of the equation by } ds$$

$$\int_0^{s_{max}} f_{iy} \cdot ds = k \int_0^{s_{max}} s \cdot ds \qquad \text{Taking the integral of the upper and lower limits: The limit of zero is the point where one propagated group meets the surface molecule, and } s_{max} \text{ is the displacement of the surface molecules due to the energy received by the same propagated group.}$$

$$f_{iy} \cdot s_{max} = k \cdot \frac{1}{2} \left[s^2 \right]_0^{s_{max}}$$

$$f_{iy} \cdot s_{max} = \frac{1}{2} k \left[s_{max}^2 - 0 \right] = \frac{1}{2} k \cdot s_{max}^2$$

$$E_{kiy} = f_{iy} \cdot s_{max}$$

$$E_{kiy} = \frac{1}{2} k \cdot s_{max}^2 \quad \text{-------- (3-1)}$$

Where: $f_{iy} = f_i \cos \alpha$ -------- (3-2)

Then:

$$E_{kiy} = s_{max} \cdot f_i \cos \alpha = \frac{1}{2} k \cdot s_{max}^2 \quad \text{-------- (3-3)}$$

If the value of the constant k for the *spring (sp)*, see fig. 3.4, is not large enough to provide the potential energy (E_{piy}) equal to E_{kiy} from the force (f_{iy}) caused by the incoming propagated group of the incident ray, then the surface molecule for that substance will not be able to provide the force (f_{ry}) needed to reflect that kind of incident ray. As a result, this surface will be a black or transparent surface.

The frequency of the surface molecules will determine the color. Some surface molecules can synchronize with a wider range of frequencies from the incoming rays and will reflect more than one color. This can be seen in equation (2-10), which is dependent on the natural frequency of the molecules (f_m) and the frequency of the incoming propagated rays (f_p).

When the molecules on the surface of a solid or liquid cannot provide a perpendicular force (f_{ry}) that is equal to f_{iy}, then this surface will be black or transparent. For black surfaces, the molecules under the surface will not be able to resonate with the frequency from the incoming propagated ray. The yaldons in that incoming ray will be scattered among the molecules of the black substance. Then the total amount of yaldons will increase among the molecules of that substance than the conservation of momentum will allow in the universe (M_y). As a result, the extra amount of yaldons will be radiated as heat as the traveling groups bombard the black substance's molecules. In the case of transparent substances, the molecules under the surface will be able to resonate with the frequency from the incoming propagated ray. Then the molecules of the transparent substance will be able to propagate the incoming light ray to the opposite side of that transparent substance. The transparent substance will be considered as a conductor for the propagated groups of visible light, but the propagated groups of infrared will not conduct with the molecules of a transparent substance in the same manner as visible light. Then this substance will be considered as a semiconductor for infrared (refer to equation 2-10).

The Refraction of Propagated Rays

The surface molecules of transparent substances will not be able to provide a force that is equal to f_{iy}. Then the force f_{iy} will be added to the force f_{py}, resulting in f_{pi} as shown in fig. 3.6 below:

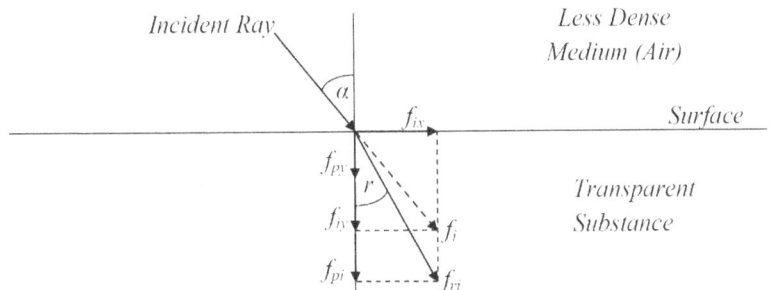

*Fig. 3.6 A rough diagram representing the angle of refraction (r) in a transparent substance.

The angle r in the figure above depends on the force f_{py}, and f_{py} depends on the size of the surface molecules (a_s will be the cross-sectional surface area of a surface molecule see fig 3.2 and 3.3). According to fig. 3.6:

$$\sin r = \frac{f_{ix}}{f_{ri}}$$

Then the sin r will be inversely proportional to f_{ri}:

$$\sin r \propto \frac{1}{f_{ri}}$$

Where: $f_{ri} \propto f_{py}$

And: $f_{py} \propto r_a^2$

Then:

$$f_{ri} \propto r_a^2$$

After substituting the proportional value of f_{ri} into the relation $\sin r \propto \dfrac{1}{f_{ri}}$:

$$\sin r \propto \frac{1}{r_a^2} \quad \text{------- (3-4)}$$

According to equation (2-9) (the velocity of a propagated group in a transparent substance):

$$v_t = \frac{v_p}{v_m} \cdot \frac{m_g}{2\pi \cdot r_a^2 \cdot \sigma \cdot m_y} \cdot \sqrt{\frac{k}{m}}$$

After rearranging the variables:

$$v_t = \frac{1}{r_a^2} \cdot \frac{v_p}{v_m} \cdot \frac{m_g}{2\pi \cdot \sigma \cdot m_y} \cdot \sqrt{\frac{k}{m}}$$

According to the equation above, the velocity of light in a transparent substance will be inversely proportional to the size of the surface molecules. Then:

$$v_t \propto \frac{1}{r_a^2} \quad \text{------- (3-5)}$$

After substituting the value of $\left(\frac{1}{r_a^2}\right)$ from relation (3-5) into (3-4):

$$\sin r \propto v_t$$

According to the relation above, the velocity of the propagated ray in a transparent substance will be proportional to the angle of refraction (r).

The surface molecule will follow the direction of the resultant force, f_{ri}, while oscillating about its rest point, and the propagated ray will move in the same direction of that force (f_{ri}, see fig. 3.6). The traveling groups of yaldons will apply a balanced force on all the molecules of that substance (except for the molecules located on the surface and near the corners) since the rectangular transparent substance has parallel sides. Then the emitted yaldon group will keep moving in a straight line within that substance (since there is no force component of f_{py} on the internal molecules) until it reaches the opposite side, where it will come into contact with another surface molecule. That surface molecule from the opposite side, as seen in fig. 3.7 below, will be under the force f_{ro}. The force f_{ro} has two components, f_{oy} and f_{ox}. The force f_{po} will equal f_{oy} minus f_{py}, and the resultant force (f_o) will be the final force with an angle of β. The angle of β will be the same angle as α from fig. 3.6 and fig.3.7.

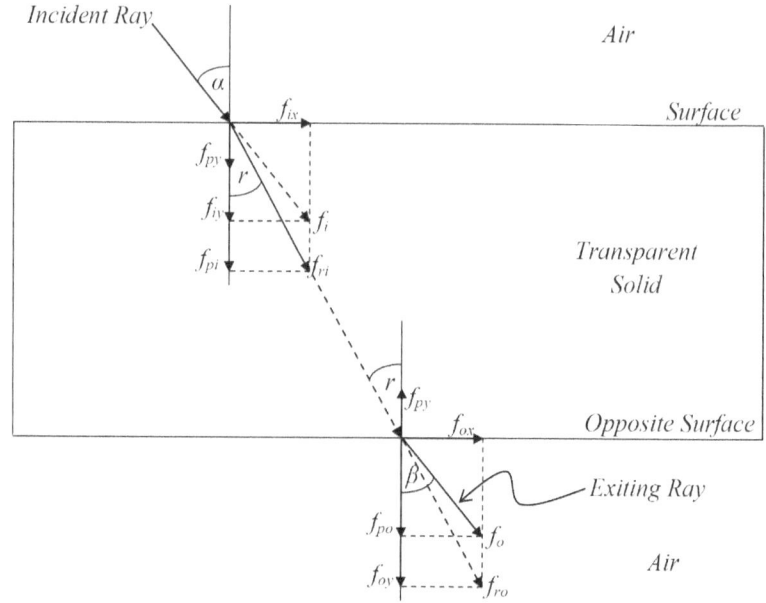

*Fig. 3.7 A rough diagram representing the propagated rays passing through a transparent solid substance and their force components on the surface molecules.

The force f_{ro} equals the force f_{ri} since the transparent solid substance will maintain the momentum of the propagated ray through all its internal molecules. According to the figure above, f_{ox} equals f_{ix}, and f_{po} equals f_{iy} since there is a force, f_{py}, that has the same value applied onto both surfaces. Then the tangent of β equals the tangent of α. As a result, the leaving propagated ray will be parallel to the incident propagated ray.

Total Reflection for Black and Transparent Surfaces

The black and transparent substance surface molecules cannot store enough of the potential energy (E_{piy}) received from the incident propagated ray in order to be reflected back to the same medium that it was received from, but equation (3-3) shows that the value of the kinetic energy (E_{kiy}) can vary with the angle of α when the constant k is fixed (for the same substance as in fig. 3.5). If the angle of α is large enough to lower the value of the kinetic energy (E_{kiy}) to a value where the constant k can store an equal amount of potential energy (E_{piy}), then the surface molecules will be able to provide the force f_{ry}, as in fig. 3.4. As a result, this substance will be able to totally reflect the incident propagated ray, as well as the traveling groups that meet its surface molecules at a large enough value for angle α. Internal reflection inside transparent substances will follow the same principle, and this gives the fiber optic wire its characteristic to transfer visible light through its length due to its conductivity for visible light.

The Separation of Colors (Dispersion)

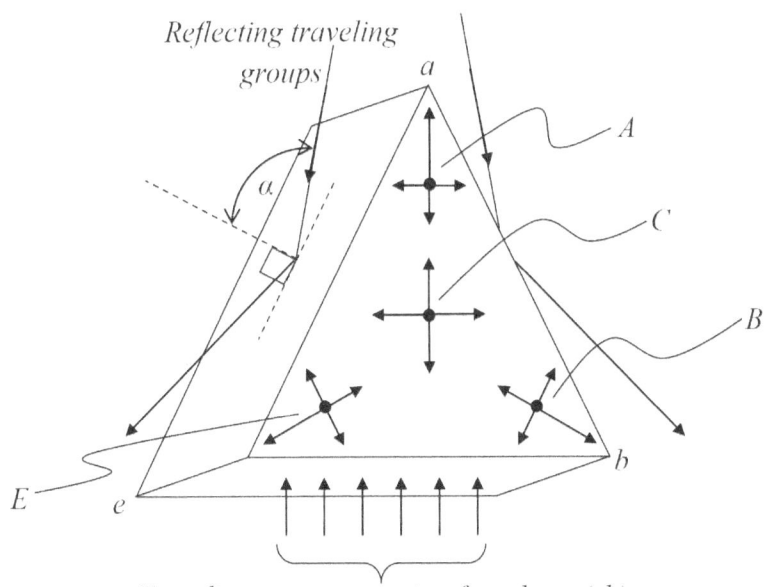

*Fig. 3.8 A rough diagram representing the uneven forces on the molecules at the corners of the prism, due to the traveling groups entering from the base and reflecting off of the sides.

The traveling groups of yaldons cause the molecules in substances to vibrate. When the transparent object is in the shape of a prism, see fig. 3.8, the molecules at the points of A, B, and E don't have an equal amount of traveling groups bombarding them from all sides. For example, point A will be struck by a less number of traveling groups from the top corner (a), since there will be a total reflection for some of the traveling groups which try to enter the prism from the areas that are close to the corner (a). As some of the traveling groups totally reflect from that corner, there will also be a large number of traveling groups that will enter the prism from the base (eb) and strike the molecules at point A. The molecules at point A will receive a greater amount of traveling groups from the base (eb) than the area close to corner (a). Then the average forces on the molecules at point A from the base (eb) will be more than the forces from the top corner (a).

As a result, the molecules at point A will vibrate with a larger span toward corner (a) than toward the base (eb) in respect to the molecules' rest point. The same principle will apply to the molecules at points B and E, but not at point C. The molecules at point C (the center of the prism) will have a more uniform oscillation around its rest point, since those molecules will have equal forces applied onto them from all directions. The molecules that are located close to the corners of the prism will have a resultant force (f_{ga}) that is pointed toward the direction of the corner that is nearest to its location, see fig. 3.9 below.

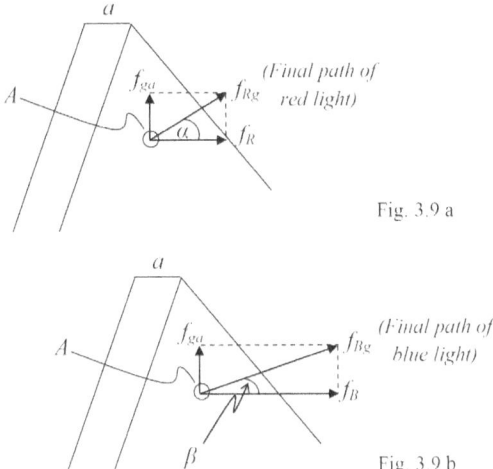

*Fig. 3.9 A rough diagram representing the angles of the red and blue light as they disperse through a transparent prism.

As the red light strikes the molecules at point A, the force of the red light (f_R) will be smaller than the force of blue light (f_B) since the total momentum of the propagated red groups of light will be lower than the total momentum of the propagated blue groups (see fig 3.9 a and b). As a result, angle α will be greater than the angle β, and the red light will appear to bend less inside the prism than the blue light. In other words, the red light will be raised higher than the orange, the orange will be raised higher than the yellow, the yellow higher than the green, and so on until the full spectrum of visible light becomes analyzed due to the uneven forces on the molecules at the corners of the prism (see figure 3.10 below).

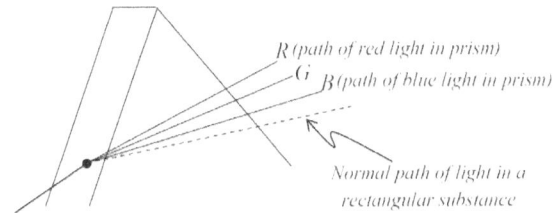

*Fig. 3.10 A rough diagram representing the path of the red light being raised more than the blue light as they travel through a transparent prism.

Diffraction

It is now a well-known and familiar concept that the surface molecules for any substance will vibrate vertically in respect to that surface. In this case, any propagated ray group that touches the surface molecules on the edge of a smooth substance tangentially (consider the angle of the incident ray, α, equal to 90° as in fig 3.4) will be diffracted away from that edge, as the ray comes into contact with the molecules that oscillates perpendicularly on the surface of that edge (see *edge-a* in fig. 3.11 and fig. 3.12). Should propagated rays go through a narrow slit, then the rays will be diffracted away from the surfaces of both edges (see *slit-b* in fig. 3.11).

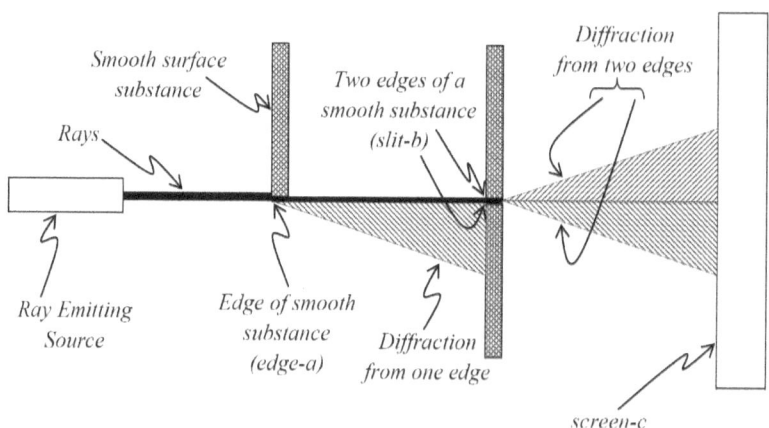

*Fig. 3.11 A rough diagram representing the diffraction of propagated rays as they come into contact with the molecules on the edge (edge-a) and the edges (slit-b) of a smooth substance.

The vertical oscillation of the molecules on the surface of *edge-a* will diffract the incoming propagated rays at varying angles. The angle of diffraction will depend upon the molecules' position from rest point as these incoming propagated rays, which are traveling tangential to the surface of *edge-a*, comes into contact with them (see the following fig. 3.12). The value of angles α_1

and α_2 will depend on the momentum of the propagated group, as well as the position from rest point of the vibrating molecules on the surface of *edge-a*. These properties will give the projection of propagated groups onto *screen-c* a random appearance.

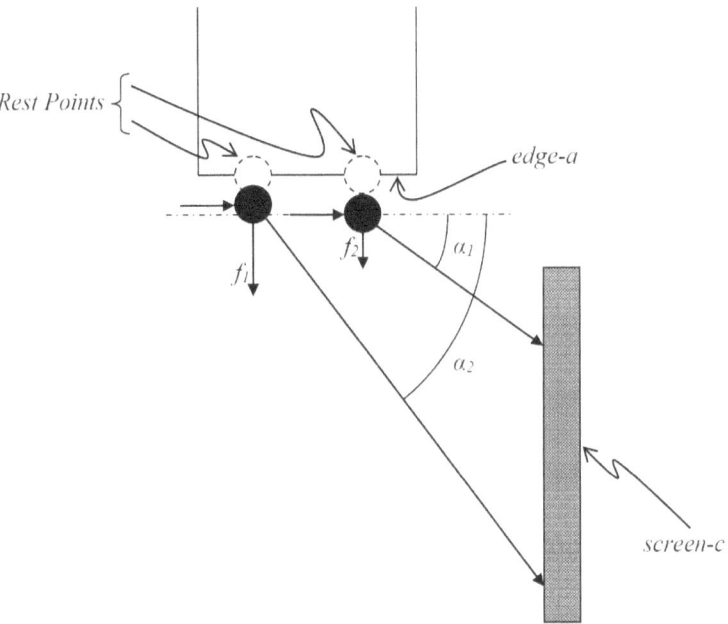

*Fig. 3.12 A rough diagram representing the randomized angles of diffraction as a propagated group comes into contact with a molecule on edge-a in regard to its position from rest point.

Double-Slit Experiment

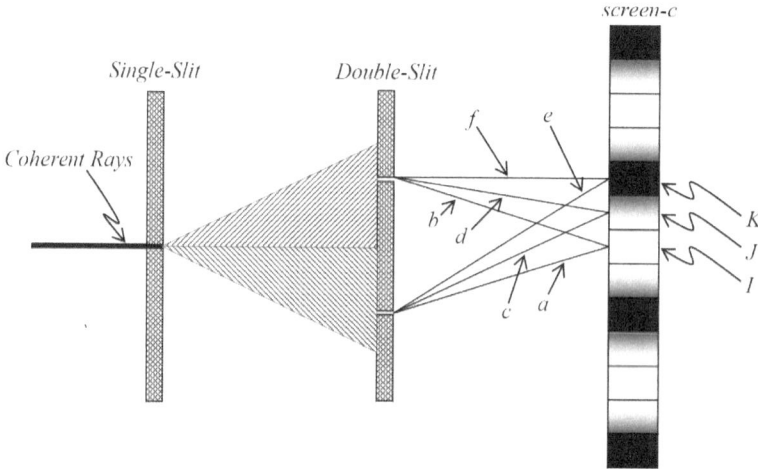

*Fig. 3.13 A rough diagram representing the different distribution
of forces onto the surface molecules of *screen-c* as propagated
rays from a coherent source pass through two slits.

In the figure above, the propagated groups in the path of rays *a* and *b* will travel the same distance to *screen-c* since these propagated groups are emitted from a coherent source. Then the propagated groups in rays *a* and *b* will meet simultaneously and strike the surface molecules of area *I* on *screen-c* with a maximum possible force. The propagated groups in the path of rays *c* and *d* will strike the surface molecules of area *J* with force that is less than the previous example of the groups which strike area *I*. This is due to the propagated groups of yaldons in rays *c* and *d* striking the area *J* with a slight time-shift. This time shift between the propagated groups will gradually increase, until the molecules on the surface of area *K* will be struck by the propagated groups in rays *f* and *e* which will have the greatest time shift possible (applying the least amount of force upon the surface molecules). This will cause area *K* to appear darker. The following equations and diagrams will validate the different distribution of forces onto *screen-c* with a relation between the resultant force, *F(P)*, the position of point *P* from the centerline, as in fig. 3.14.

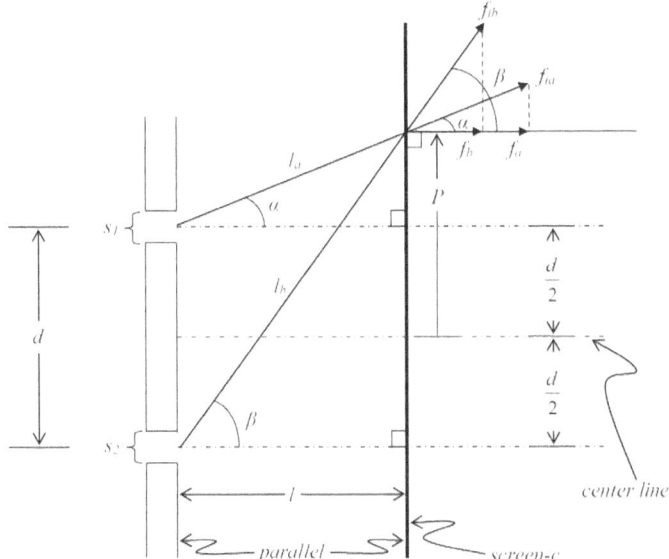

*Fig. 3.14 A rough diagram representing the different forces onto the surface molecules of *screen-c* at the height of *P* from center line as propagated rays pass through two slits.

The force component (f_a) from the propagated group of the ray (l_a) which passes through the slit (s_1) will be applied perpendicular component onto the surface molecules of *screen-c* at the height of *P* from the centerline (see fig. 3.14 above). This force can be represented by half of a cosine wave, as shown in the following fig. 3.15 and equation (3-6).

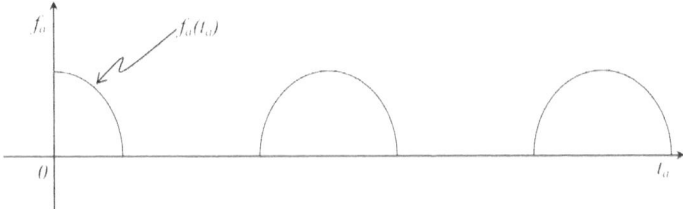

*Fig. 3.15 A rough diagram representing the force component f_a that is caused by the propagated groups from *ray-a* onto ¬screen-c ¬as a function of t_a with a half cosine waveform.

The function $f_a(t_a)$ can be written as follows:

$$f_a\left(t_a\right)=0.5\left[f_{ia}\cdot\cos\ \left(2\pi\cdot f_c\cdot t_a\right)+\left|f_{ia}\cdot\cos\ \left(2\pi\cdot f_c\cdot t_a\right)\right|\ \right]-------(3\text{-}6)$$

Where: f_c is the frequency of the coherent rays that travel the path of l_a and l_b. The surface molecules on *screen-c* at the height of P are also under another perpendicular force component $(f_b(t_b))$ from the ray that travels the path of l_b through the second slit, s_2. Then the resultant force on the surface molecules at the height of P will be $F_d(t)$:

$$F_d\left(t\right)=f_a\left(t_a\right)+f_b\left(t_b\right)$$

$$F_d\left(t\right)=0.5\Big[\big[f_{ia}\cdot\cos\ \left(2\pi\cdot f_c\cdot t_a\right)+f_{ib}\cdot\cos\ \left(2\pi\cdot f_c\cdot t_b\right)\big]+$$

$$\left|f_{ia}\cdot\cos\ \left(2\pi\cdot f_c\cdot t_a\right)+f_{ib}\cdot\cos\ \left(2\pi\cdot f_c\cdot t_b\right)\right|\ \Big]-----(3\text{-}7)$$

Where:

$$f_c=\frac{v_p}{\lambda_c}$$

And:

$$t_a=\frac{l_a}{v_p}\ \text{also}\ t_b=\frac{l_b}{v_p}$$

After substituting the values of f_c, t_a, and t_b into equation (3-7):

$$F_d\left(P\right)=0.5\left[\left[f_{ia}\cdot\cos\ \left(\frac{2\pi}{\lambda_c}\cdot l_a\right)+f_{ib}\cdot\cos\ \left(\frac{2\pi}{\lambda_c}\cdot l_b\right)\right]+\left|f_{ia}\cdot\cos\ \left(\frac{2\pi}{\lambda_c}\cdot l_a\right)+f_{ib}\cdot\cos\ \left(\frac{2\pi}{\lambda_c}\cdot l_b\right)\right|\right]$$

$-----(3\text{-}8)$

From fig. 3.14:

$$l_a=\sqrt{\left(P-\frac{d}{2}\right)^2+l^2}$$

$$l_b = \sqrt{\left(P + \frac{d}{2}\right)^2 + l^2}$$

$f_{ia} = f_{ib} = f_i$ Coherent ray source (propagated groups have the same momentum)

$$f_a = f_i \cos\alpha = f_i\left(\frac{l}{l_a}\right)$$

$$f_a = f_i \frac{l}{\sqrt{\left(P - \frac{d}{2}\right)^2 + l^2}}$$

In a similar way for f_b:

$$f_b = f_i \cos\beta = f_i\left(\frac{l}{l_b}\right)$$

$$f_b = f_i \frac{l}{\sqrt{\left(P + \frac{d}{2}\right)^2 + l^2}}$$

After substituting the values of l_a, l_b, f_a, and f_b into equation (3-8):

$$F_d(P) = 0.5 f_i \cdot l \left[\left[\frac{\cos\left(\frac{2\pi}{\lambda_c} \cdot \sqrt{\left(P - \frac{d}{2}\right)^2 + l^2}\right)}{\sqrt{\left(P - \frac{d}{2}\right)^2 + l^2}} + \frac{\cos\left(\frac{2\pi}{\lambda_c} \cdot \sqrt{\left(P + \frac{d}{2}\right)^2 + l^2}\right)}{\sqrt{\left(P + \frac{d}{2}\right)^2 + l^2}}\right] + \right.$$

$$\left. \left|\frac{\cos\left(\frac{2\pi}{\lambda_c} \cdot \sqrt{\left(P - \frac{d}{2}\right)^2 + l^2}\right)}{\sqrt{\left(P - \frac{d}{2}\right)^2 + l^2}} + \frac{\cos\left(\frac{2\pi}{\lambda_c} \cdot \sqrt{\left(P + \frac{d}{2}\right)^2 + l^2}\right)}{\sqrt{\left(P + \frac{d}{2}\right)^2 + l^2}}\right|\right] \quad \text{----- (3-9)}$$

Formula for Multi-Slit Experiment (Diffraction Grating)

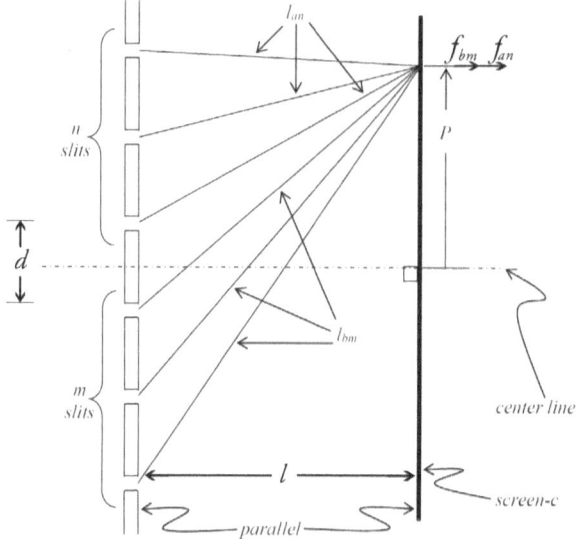

*Fig. 3.16 A rough diagram representing the forces onto the surface molecules of *screen-c* at the height of *P* as propagated rays from a coherent source pass through multiple slits.

The surface molecules on *screen-c* at the height of *P* from the centerline will be under a multitude of different forces from the propagated groups from the rays that travel through the multiple slits (as in fig. 3.16). To find the net force applied onto the surface molecules at the height of *P*, a summation will be performed on all the forces caused by the rays in l_{an} and l_{bm}.

The net force will be $F_{dg}(P)$:

$$F_{dg}(P) = \sum_{n=1}^{N} F_{an}(P) + \sum_{m=1}^{M} F_{bm}(P) \quad \text{- - - - - (3-10)}$$

Where:

$$l_{an} = \sqrt{\left(P - \left(d(n-1) + \frac{d}{2}\right)\right)^2 + l^2}$$

$$l_{hm} = \sqrt{\left(P + \left(d(m-1) + \frac{d}{2}\right)\right)^2 + l^2}$$

After rewriting equation (3-10):

$$F_{dg}(P) = 0.5 \left[\left[\sum_{n=1}^{N} f_{an} \cdot \cos\left(\frac{2\pi}{\lambda_c} l_{an}\right) + \sum_{m=1}^{M} f_{bm} \cdot \cos\left(\frac{2\pi}{\lambda_c} l_{bm}\right)\right] + \right.$$

$$\left.\left|\sum_{n=1}^{N} f_{an} \cdot \cos\left(\frac{2\pi}{\lambda_c} l_{an}\right) + \sum_{m=1}^{M} f_{bm} \cdot \cos\left(\frac{2\pi}{\lambda_c} l_{bm}\right)\right|\right] - - - - - (3\text{-}11)$$

Where N is the number of the slits above the center line and M is the number of slits below that same centerline (as in fig. 3.16). Then the forces applied onto the surface molecules at the height of P, due to the coherent ray source going through the multiple slits, will be f_{an} and f_{bm}:

$$f_{an} = f_i \frac{l}{l_{an}}$$

$$f_{bm} = f_i \frac{l}{l_{bm}}$$

Where: f_i is the force from the coherent ray source that passes through any of the slits.

After substituting l_{an}, l_{bm}, f_{an}, and f_{bm} into equation (3-11):

$$F_{dg}(P) = 0.5 f_i \cdot l \left[\left[\sum_{n=1}^{N} \frac{\cos\left(\frac{2\pi}{\lambda_c} \sqrt{\left(P - \left(d(n-1) + \frac{d}{2}\right)\right)^2 + l^2}\right)}{\sqrt{\left(P - \left(d(n-1) + \frac{d}{2}\right)\right)^2 + l^2}}\right. + \right.$$

$$\sum_{m=1}^{M} \frac{\cos\left(\frac{2\pi}{\lambda_c}\sqrt{\left(P+\left(d(m-1)+\frac{d}{2}\right)\right)^2+l^2}\right)}{\sqrt{\left(P+\left(d(m-1)+\frac{d}{2}\right)\right)^2+l^2}} +$$

$$\left|\sum_{n=1}^{N} \frac{\cos\left(\frac{2\pi}{\lambda_c}\sqrt{\left(P-\left(d(n-1)+\frac{d}{2}\right)\right)^2+l^2}\right)}{\sqrt{\left(P-\left(d(n-1)+\frac{d}{2}\right)\right)^2+l^2}} +\right.$$

$$\left.\sum_{m=1}^{M} \frac{\cos\left(\frac{2\pi}{\lambda_c}\sqrt{\left(P+\left(d(m-1)+\frac{d}{2}\right)\right)^2+l^2}\right)}{\sqrt{\left(P+\left(d(m-1)+\frac{d}{2}\right)\right)^2+l^2}}\right| \quad - - - - - (3\text{-}12)$$

Polarization

When the propagated visible light groups come into contact with the molecules of a transparent substance, the molecules of the transparent substance will emit propagated groups from the opposite side of where the contact was made by the propagated visible light groups. The emitted propagated group from the molecules will take the shape from the molecules' structure. For example, if the shape of the molecule is spherical, then the emitted propagated group will take a spherical shape (like a round shell). In the case of a polarized substance, the shape of the molecule structure is rectangular with a larger length than its width. Then the shape of the propagated ray will also be rectangular (see fig. 3.17).

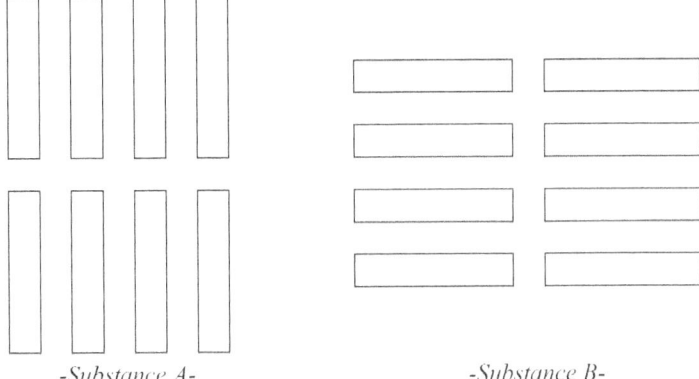

-Substance A-   -Substance B-

*Fig. 3.17 A rough diagram representing the shape of the emitted propagated ray from two different substances' polarized molecules.

From the figure above, the emitted propagated groups from *substance A* can't make the molecules in *substance B* vibrate with a large enough distance in order to replicate the momentum from the incoming propagated rays. As a result, no light will be able to pass through *substance B* from *substance A*.

Greenhouse Effect

The greenhouse effect that contains heat within an enclosed area is caused by certain substances being able to transfer light, but they cannot transfer infrared at the same rate. For example, a typical greenhouse is an enclosed room with walls made from glass. The glass molecules will resonate with the light rays; thus it will be considered as a conductor for light. But the glass molecules will not resonate with infrared. As a result, glass will be considered as a semi-conductor (to a certain degree) for the propagated groups of infrared (heat). When light enters through the walls of a greenhouse, it will fall on different items within that greenhouse. These items will have different colors for their surfaces and will reflect some of the light back through the glass walls of the greenhouse. The light that was not reflected will be absorbed among the items as yaldon particles. As mentioned earlier in chapter 1, this will increase the amount of yaldons among the molecules of the items inside the greenhouse more than the conservation of momentum will allow. Due to the continuous bombardment of the traveling groups of yaldons onto the molecules of the items in the greenhouse, the extra yaldon particles will be released as propagated groups of infrared. This released infrared from all the items in the greenhouse will not be able to transfer through the glass walls (leave the greenhouse at the same rate as visible light), since glass is a semi-conductor for infrared. As a result, the rate of propagated groups of yaldons entering the greenhouse will be more than the propagated groups that leave the greenhouse as the sunlight enters through the glass. The same effect will apply to a parked car in direct sunlight. To keep the car as cool as possible is to increase the amount of yaldon particles that can leave the car. This is done by using surfaces that will reflect the most visible light, like a smooth white or silver reflecting surface placed close to the windshield. This will reflect back as much yaldons as possible through the windshield and out of the car.

Electric Current

The electric current that is generated inside of electric conductors happens in a similar fashion to the way that a laser is generated inside of a transparent substance; both require an external force to agitate and supply their molecules and atoms with yaldon particles. Also, they both are usually surrounded by air (air is considered an insulator for an electric current but a conductor for the light). Since it is known that the light from a laser is generated by the simple harmonic motion of the molecules in a transparent substance, then the electric current is generated by the vibration of the atoms in a metal substance, as a series of several groups of a propagated ray of yaldons, with a frequency governed by the simple harmonic motion of those same atoms in the metal substance. In this way, the electric current can be considered as a propagated ray. Then the electric current will be referred to as an *electric ray*.

This will be able to explain the reason for an electric ray to diffract as it passes through a double-slit and strikes a screen, since the electric ray that passes through the slits is generated from a coherent source; just like the light and both are comprised of a group of yaldons. In the diffraction experiment using an electric ray, it had been noticed that there will be no diffraction for the electric ray on the screen when using a measuring device. Using any kind of measuring device to sense a group of yaldons will change its frequency (λ_c as in equation 3-9), and this will cause the propagated groups of the electric ray that passes through the slits (s_1 and s_2 as in fig. 3.14) to no longer be from a coherent source. As a result, there will be no interference pattern onto the projected screen. In other words, if λ_c is altered by a very small fraction in one of the two slits, then there will be no interference fringes on the screen.

CHAPTER 4

———◆—‹•••›◆‹•••›———◆———

The Periodic Table of the Elements

The Atomic Structure of the Periodic Table of the Elements According to the Yaldon Model

This will be a brief discussion about the shape of the atoms according to the periodic table of elements without going into detail on the nature of the forces applied on and by the atoms. The shape of the hydrogen atom is similar to the shape of a galaxy, disk-like with a dense mass in the center. The dense mass in the center of the hydrogen atom will spin rapidly about its axis, and the yaldon particles that surround it will whirl about the dense core, creating a disk-like shape. The conservation of momentum in the universe will maintain a steady amount of yaldon particles that move around the dense spinning core. When there is a steady amount of yaldon particles around the core, then this will be known as a neutral element (neutron). Should there be a shortage of yaldons around this dense core, then that atom will be positive. This is due to the movement of the yaldon particles *toward* the spinning dense core. If there is an extra amount of yaldons around the spinning core, then the conservation of momentum in the universe will move the extra yaldon particles away from the spinning dense core. When the extra yaldons move *away* from the dense core, this atom will appear to have a negative charge. Since the spinning dense core of mass, without the yaldons whirling around it, appears to be positively charged, then this rapidly spinning dense core of mass will be called a proton. Following will be fig. 4.1 representing a galactic model for the hydrogen atom.

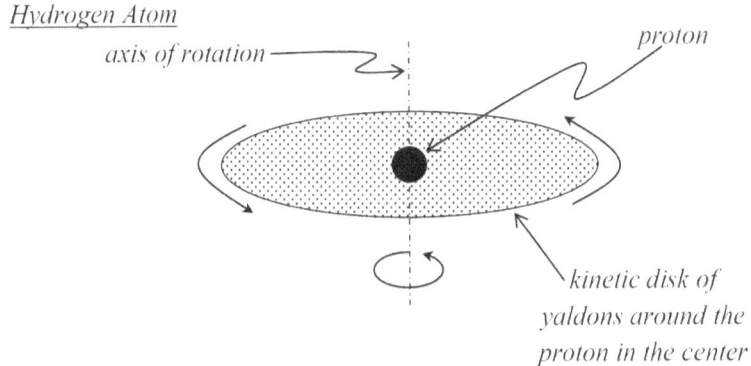

*Fig. 4.1 A rough diagram representing the galactic model of a hydrogen atom. This will be the basic building block for the rest of the elements

The next element in the periodic table is helium. Helium will consist of a central shaft of two hydrogen atoms that are stacked on top of each other with a short distance between the two atoms (not touching each other) and another two hydrogen atoms which revolve on a plane perpendicular to the central shaft (forming *orbit-a*). The central shaft of hydrogen atoms will create a suction force perpendicular to its axis of rotation. The central shaft, along with the two orbiting atoms, will create a gyroscopic model. The central shaft can be considered as a nucleus of an atom. The rest of the elements in the periodic table will also be gyroscopic. Following will be fig. 4.2, which will model this concept.

Helium (He)

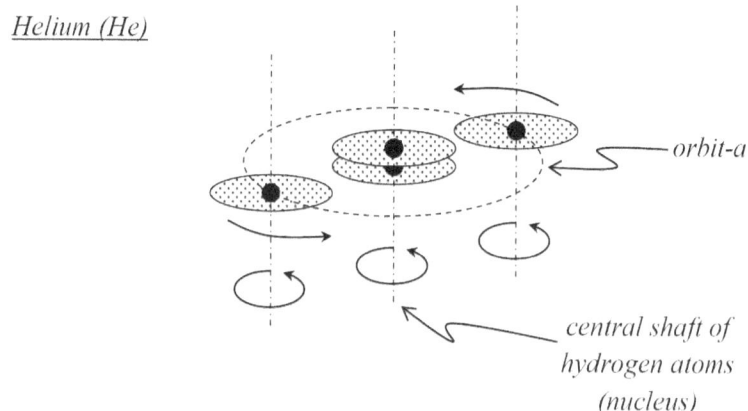

orbit-a

central shaft of
hydrogen atoms
(nucleus)

*Fig. 4.2 A rough diagram representing the gyroscopic model of
a helium atom, with two hydrogen atoms that fill *orbit-a*.

The next atom on the periodic table of elements is lithium. Lithium will require a new orbit, *orbit-b*. Following will be fig. 4.3, which will model the lithium atom and *orbit-b*.

Lithium (Li)

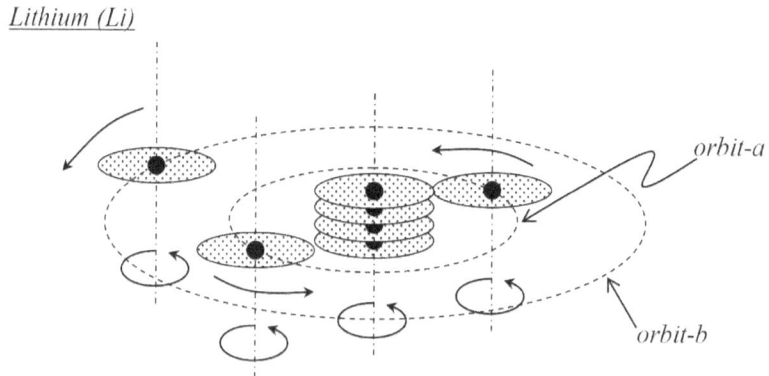

orbit-a

orbit-b

*Fig. 4.3 A rough diagram representing the model of a lithium
atom, with four hydrogen atoms in the central shaft. There will
be two hydrogen atoms in *orbit-a,* and one in *orbit-b*.

The newly added *orbit-b* will be able to hold another seven hydrogen atoms. This will be able to fulfill the second row of the periodic table of elements up to neon. The central shaft (nucleus) of the atom will continue to collect hydrogen atoms until the neon atom will have ten hydrogen atoms in its nucleus. The neon atom will also have two hydrogen atoms in its inner orbit (*orbit-a*) and eight hydrogen atoms in its outer orbit (*orbit-b*). Following will be a rough diagram representing the general shape of a neon atom (fig. 4.4).

Neon (Ne)

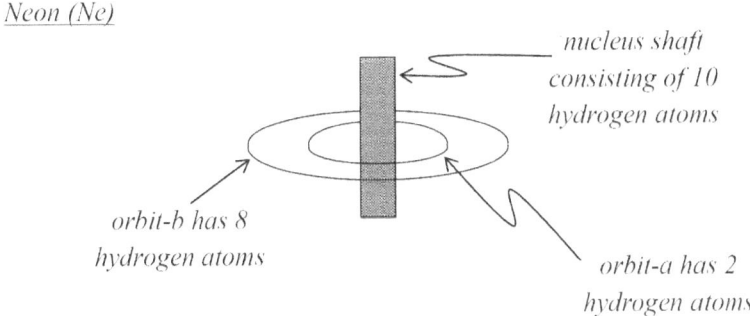

nucleus shaft consisting of 10 hydrogen atoms

orbit-b has 8 hydrogen atoms

orbit-a has 2 hydrogen atoms

*Fig. 4.4 A rough diagram representing the shape of a neon atom.

For the atoms in the third row of the periodic table of elements from sodium (Na) to argon (Ar), they will form another orbit (*orbit-b'*) parallel to *orbit-b*. *Orbit-b'* will be able to have another eight hydrogen atoms in its orbit, and this will be able to fulfill the periodic table of elements from the sodium atom to the argon atom. The nucleus shaft will also continue to stack hydrogen atoms onto it, growing in height as the atoms grow heavier. The force of the nucleus shaft that is perpendicular to the axis of rotation, in the direction toward the nucleus, will also increase as the nucleus's height increases. Following will be fig. 4.5 modeling the general shape of an argon atom.

Argon (Ar)

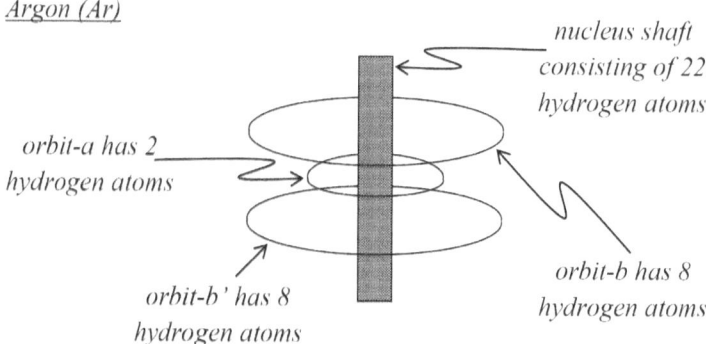

nucleus shaft consisting of 22 hydrogen atoms

orbit-a has 2 hydrogen atoms

orbit-b' has 8 hydrogen atoms

orbit-b has 8 hydrogen atoms

*Fig. 4.5 A rough diagram representing the shape of an argon atom.

The atoms in the fourth row of the periodic table of elements from potassium (K) to krypton (Kr) will form a new ring for *orbit-c. Orbit-c* can hold a maximum number of eighteen hydrogen atoms in its orbital path. Following will be fig. 4.6, modeling the shape for the atoms from potassium to krypton in the periodic table of elements.

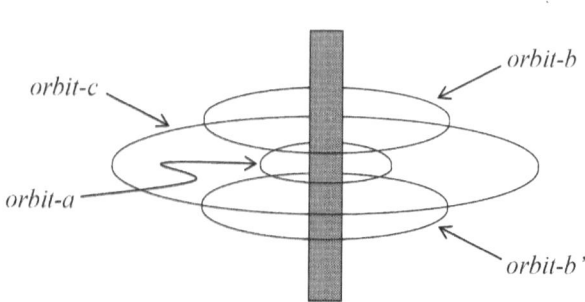

orbit-c

orbit-a

orbit-b

orbit-b'

*Fig. 4.6 A rough diagram representing the shape of the atoms
in the periodic table of elements from potassium (K) to krypton
(Kr). The growing nucleus shaft will have a range of 20 to 48
hydrogen atoms within it; from potassium to krypton.

The atoms in the fifth row of the periodic table of elements from rubidium (Rb) to xenon (Xe) will form a fifth orbital ring (*orbit-c'*) that can also hold eighteen hydrogen atoms and will be parallel to *orbit-c* (see the following fig. 4.7).

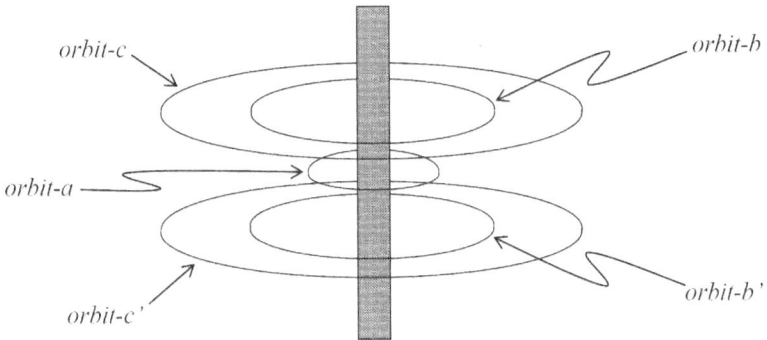

*Fig. 4.7 A rough diagram representing the shape of the atoms
in the periodic table of elements from rubidium (Rb) to xenon
(Xe). The growing nucleus shaft will have a range of 48 to
77 hydrogen atoms within it; from rubidium to xenon.

The atoms cesium (Cs), barium (Ba), and lanthanum (La)
will have hydrogen atoms that occupy a sixth orbit (*orbit-c "*). The
first six atoms in the lanthanide series, from cerium (Ce) to euro-
pium (Eu), will expand *orbit-a*. The path of *orbit-a* will increase in
size and hold six more hydrogen atoms for a total of eight hydrogen atoms in its orbit. The name of *orbit-a* will change to *orbit-b "*
after the increase in size. The rest of the atoms in the lanthanide
series, from gadolinium (Gd) to lutetium (Lu), will form stacks
of two hydrogen atoms in *orbit-b "* for a total of sixteen hydrogen
atoms (eight stacks of two hydrogen atoms). The following atoms
in row six of the periodic table of elements, from hafnium (Hf)
to radon (Rn), will fill *orbit-c "* and add another fifteen hydrogen
atoms to this orbit (eighteen total hydrogen atoms in *orbit-c "* for
radon). Following will be fig. 4.8, which will show the general
shape of the atoms in row six of the periodic table of elements. In
fig. 4.8, the bold line for the obit path of *orbit-b "* represents the
stacks of two hydrogen atoms that occupy that path.

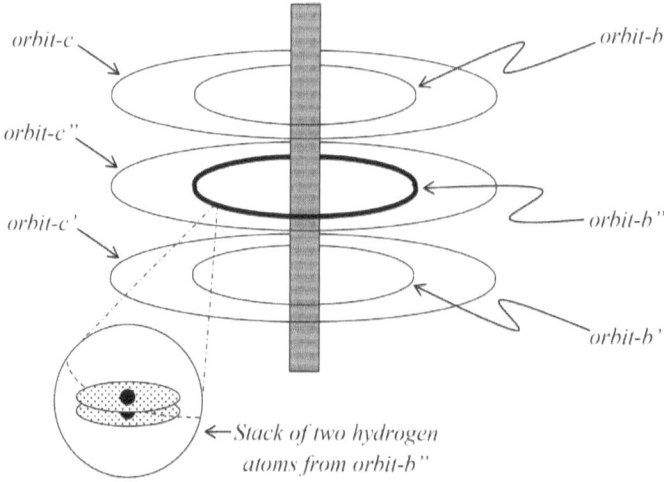

*Fig. 4.8 A rough diagram representing the shape of the atoms in the periodic table of elements from cesium (Cs) to radon (Rn). The growing nucleus shaft will have a range of 78 to 136 hydrogen atoms within it; from cesium to radon.

The atoms in the seventh row of the periodic table of elements, from francium (Fr) to rutherfordium (Rf) and including all the atoms in the actinide series, will form stacks of two hydrogen atoms in *orbit-c"*. Fig. 4.9 will be a rough diagram modeling this concept with a bold line for *orbit-c"*.

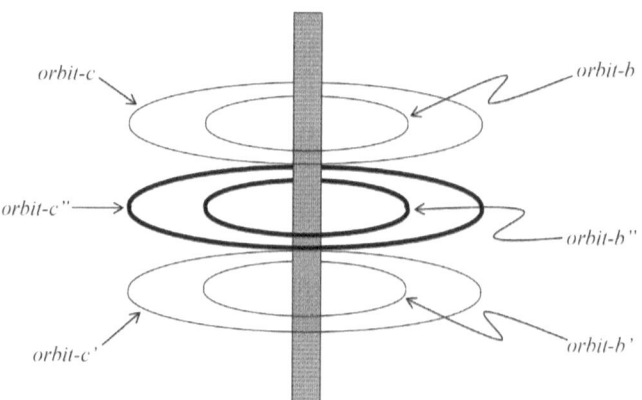

*Fig. 4.9 A rough diagram representing the shape of the atoms in the periodic table of elements from francium (Fr) to rutherfordium

(Rf). The growing nucleus shaft will have a range of 136 to 157
hydrogen atoms within it; from francium to rutherfordium.

The rest of the synthetic elements in the seventh row of the periodic table of elements, starting from dubnium (Db), will form stacks of two hydrogen atoms in *orbit-b*, *b'*, *c*, or *c'*. In this way, the atomic model of the periodic table of elements can be fulfilled with a gyroscopic model of a central rotating shaft with orbital rings that form a well-distributed area for the mass contained within the atom.

This model of yaldon particles was developed since, according to Newton's second law of motion ($F\,t = m\,v$), there cannot be momentum, force, energy, or time without a mass in a bounded system with relative motion to other points of references in that system. All of the above formulas, discussions, and diagrams are used to implement this principle on the atomic scale of the universe without having to rely on an alternate reality. The term *yaldon* was chosen for the assumed particle since the root of the word, *yalda*, is translated into winter solstice from Aramaic. Winter solstice marks the first day of the year when the daylight begins to become longer, and these yaldons are responsible for light. Hopefully, this model will be used to help humanity further understand the world and the universe which it resides, in order to develop helpful tools and devices to promote a harmonious world for all.

YALDON THEORY

This model, the Yaldon theory, is the only theory that is capable to explain all of the phenomena in the physical realm with the use of only one assumption. No other theory can work as well to explain these phenomena. Examples of some of the phenomena explained by the Yaldon theory are

1. Light's peculiar behavior as both a particle and wave.
2. The constant oscillation of all the atoms and the molecules in the universe.
3. The constant emission of infrared rays from atoms.
4. The release of heat from sugar being metabolized in living bodies.
5. The cosmological redshift from the stars of the universe.
6. Find the maximum distance allowed for information to be received from an observer from space.
7. The change in the speed of light as it goes through transparent substances.
8. The spinning of the fans in a radiometer and the emission and absorption lines in lowpressure gases.
9. Light's constant speed in empty space.
10. Solid objects not being able to be placed backed together once being broken apart.
11. No interference pattern for the electron ray when using a measuring device in the double-slit experiment.
12. The charge in atomic particles.

ABOUT THE AUTHOR

Menketh Yalda's debut novel, *The Yaldon Particle Theory*, was inspired by his beliefs in Newton's model of light. In high school, he studied physics which gave him the background and inspiration to build up the yaldon theory all throughout his years at Basrah University, which he graduated in 1976 as an electrical engineer. He worked as an instructor at Basrah Electrical Institute. After this, he was sent to work at the Ratba Broadcast Station. Due to the pressure on Menketh to be part of the Baath Party, he decided to flee to Kuwait.

In July 1978, he and his wife worked in Kuwait as engineers; he was in the Ministry of Electricity in the high-tension department for eight years until 1986. During his years in Kuwait, he would spend his free time developing this theory.

In July 1986, he left Kuwait with his wife and three children and came to the United States where he pursued another bachelor's degree in computer science. He was later blessed with a fourth child, born in the United States, who then helped him to further develop this theory.

www.ingramcontent.com/pod-product-compliance
Lightning Source LLC
Chambersburg PA
CBHW021007180526
45163CB00005B/1916